BOOK 1
SPACE FRONTIERS SERIES

MW00461710

SPACE
FRONTIERS
DANGEROUS LIAISONS

Michael D'Ambrosio

Quantum
Discovery
A LITERARY AGENCY

ISBN
978-1-959314-88-2 (Paperback)
978-1-959314-89-9 (eBook)

TABLE OF CONTENTS

I

AN UNLIKELY RECRUIT

T wo alien soldiers, armed with pulse rifles and clubs, stood in a dimly lit hangar, surveying three Trident warships. Each of the Tridents had three long wands extending from its tail section in a T-shaped configuration, facing forward. The soldiers looked like knights with only their eyes showing through their helmets. One had two red stripes on his arm indicating the rank of major. The other had a gray shield with strange symbols affixed to his arm. He was an enlisted soldier named Tsig. The two were in an animated discussion.

Tsig warned, "I don't think we should trust her, Major."

Major Duowd replied, "Nonsense. Once Cherenka has obtained the remaining components for the Tridents' secret weapon, then no one will be able to stop us!"

"Conquering these infidels will be sweet," Tsig commented. "I do think we need to be wary of a trap."

Major Duowd admitted that a trap is the only thing that could possibly prevent their conquest of the galaxy. Both were eager to get revenge for their demise at the hands of the alliance.

· · · · · · ● · · · · · · · · ·

The *Luna C* glided into the canyon and through the tunnel to the underground base where it docked. Its hull looked new after repairs and resurfacing of its outer sheath and a new core had been installed.

Will and Jack approached the dock anxiously. Will wore a black, hooded cloak over a cream-colored shirt and black pants. His sword and scabbard were strapped to his waist, half-hidden by the cloak. He walked with the aid of a wooden cane.

Jack wore tight, brown, leather pants, a loose-fitting white shirt and black boots. A bandage covered his left temple. Six stitches marred his chin. His light brown hair was tied back in a small ponytail.

"They've been gone quite a while, almost two months I think," Will remarked.

"Yeah, I wondered if Maya would ever come back. This was her chance to leave me if she wanted to."

"I'm sure there's an interesting reason for her delay."

"I'll bet there is," Jack agreed.

They watched as the ship's hatch hissed and squealed. When the noise subsided, the hatch opened. Maya exited through the ship's hatch, followed by a tall, dark-skinned female officer. The sight of the officer stunned Jack and Will, leaving them mortified.

"It's not too late to run," Jack muttered.

"I can't run, you fool," Will chided. "That's why I have a cane."

"Oh, this is going to suck."

Maya called to them excitedly, "Hey, guys, look who I brought!"

Will and Jack were speechless as they gawked at Imperial General Furey, the officer behind Maya. With her hair pulled back in a tight bun and her uniform tight-fitting, she had an imposing and unemotional appearance. Will knew of her by reputation only and smart people feared her.

During Jack's short stint in the Fleet, the two had several run-ins. Furey eagerly signed Jack's separation papers to personally ensure she was rid of him. Now she stood in front of him with cold eyes staring into his soul.

"Well, Jack Fleming, it's been a long time," she commented cynically.

"Did you miss me, dear?" Jack asked with obvious sarcasm.

"Should I have?" Furey countered and then turned her attention to Will.

Will greeted her professionally, "Welcome General. Lieutenant Saris, at your service." Maya chuckled at Will's formality. Jack rolled his eyes over Will's suck up attitude.

Furey smiled at Will and remarked, "You don't look dressed for service, and your hair does seem a bit long to be regulation."

Will's eyes dropped to the ground in embarrassment. He hid the cane behind his back and responded humbly, "I...I didn't know you were coming."

Imperial General Furey inspected Will from head to toe. She noticed the sword and scabbard under his cloak. She clutched the hilt of the sword and slid it halfway out. The cross and roses on the hilt had caught her eye. She studied the hilt more closely and noticed the skull. "Interesting. So you take this pirate business seriously?"

"Excuse me, ma'am?" Will replied nervously.

"Relax, Will. Maya told me everything."

"Why don't we all go up to the hall and discuss some items of importance?" Maya suggested. Will gestured for the ladies to lead the way up the granite stairs.

Imperial General Furey looked back at Will and nodded in appreciation of the courtesy. Her eyes widened in surprise when she noticed the huge ship behind the *Luna C*. "What in the world is that?" she asked, her voice excited.

"It's the *Leviathan* - the only remaining destroyer built for the Attradeans," Will answered proudly.

"My, you are good," Furey complimented him. "I heard they were all destroyed on Ramses-3."

"Officially, they were," Will responded, wondering how much trouble they were in.

"I'd love to see her sometime."

"That can be arranged," he offered, hoping to smooth over any pending retribution.

When they reached the top of the stairs, Will took the lead and hobbled along the beautiful marble corridor to the main hall. He opened the oak doors and stood aside for the women to enter. Maya and Imperial General Furey took seats at the first glass table inside the ornate room.

Will and Jack hesitated in the doorway and glanced at each other suspiciously. "What do we do now?" Jack whispered.

"I have no idea. Of all people, why would Maya bring *her* back here?"

"Furey said she'd always be my worst nightmare. She wasn't kidding," Jack complained. Reluctantly, they entered the hall and sat down with the women.

Maya announced proudly, "Steele is joining our group."

"Who's Steele?" Jack asked, confused.

Imperial General Furey answered, "I am."

"What kind of name is Steele?"

"It's the name I was given at birth. Is there something wrong with that?"

Maya kicked Jack's shin under the table. Jack's face turned red with embarrassment. "Sorry, but you don't look like a Steele."

Furey ignored him and turned her attention to Will. "Maya warned me that the Weevil have infiltrated Galactic Security Services and possibly the Fleet itself," she revealed. "I've noticed strange behavior in some of my cabinet members and I've been able to trip up a few of them with decoy questions."

"Isn't there any way for you to weed them out?" Will inquired, curious.

"I'll be honest with you; they have too much of a head start," she admitted. "Corruption is rampant in the Fleet and, apparently, so are the Weevil."

"How could this happen under your command?" Jack taunted. "Shame on you."

Steele remained oblivious to Jack's gests and continued, "There have been numerous security leaks in the Fleet's upper echelon. We secretly moved our headquarters to Earth in the Solar System. Somehow, the Boromians were there, waiting to pounce on us. If you hadn't destroyed their armada, things would have been disastrous. I owe you my sincere thanks."

Will nodded and responded politely, "You're welcome, ma'am."

"How did you pull off such a stunt with a single ship?" she questioned him.

"Tactical flying, strategic targeting, and the element of surprise."

"I suspected someone from the Fleet had to be behind the feat. There's no way anyone else would be brazen enough to attack the Boromians, particularly through the Fleet's own portals. I just couldn't understand how it could be done with an Attradean ship. That made no sense at all."

"That was the whole point," explained Will. "No one would have reason expect it."

"When Maya mentioned that a contingent of renegade pirates was helping her, I knew something was up. There was too much cooperation and flawless execution."

Will explained, "We didn't want to break code and use a Fleet ship for an unauthorized mission so I hijacked the Attradean battle cruiser. When we made our attack on the Boromians, they thought it was King Tenemon. He's had his hands full ever since."

"Would you mind having me as a member of your crew?" Steele asked humbly.

Will was stunned and replied, "Not at all, General." This was something he never expected to hear. Jack coughed several times to indicate his displeasure.

"Please, call me Steele," she responded, "since I'm one of you, now."

"Then Steele it is," Will said.

"When is your next mission?"

"Tomorrow morning. Our target is a Ceratoan supply base in the fifth quadrant."

"I'd like to come, if you don't mind."

"It could be dangerous," Will warned.

"I assure you; I can hold my own."

Jack taunted, "I'll bet you can."

Maya slapped the back of Jack's head and chided, "Behave already, will you?"

Steele was amused. "That's okay, Jack. I can give it back, too, but without regard for protocol now. Just be careful what you start; you may not like how it ends." Jack laughed at her comment.

Maya explained meekly, "Don't mind Jack. He's been alone for too long."

"I suspected as much," she retorted. Jack frowned at them.

"Maya can show you to your quarters, if you'd like to settle in," Will suggested.

"Thank you, Will. I would like to change into something... less official."

Maya winked at Jack and Will as she and Steele left the table. Will waited until they were out of sight and complained, "What a surprise this is."

Jack grumbled, "I'd love to know what Maya was thinking."

Shanna and Celine entered the hall. Shanna wore her long, blond hair in five braided pigtails. Her mid-length dress was light blue. Celine wore black slacks and a V-neck, lavender sweater. Her hair was tied back in a stiff ponytail standing nearly upright.

Shanna exclaimed cheerfully, "There's my handsome prince!"

"I'm going to gag," Jack mumbled.

"Cool it. She'll hear you," Will whispered.

The girls sat at the table next to Will and Jack. Shanna kissed Will's cheek and teased, "Hi, baby."

Embarrassed, Will placed an arm around her and replied somberly, "Hi, Shanna."

"I see you boys have a new friend," Celine kidded. The men frowned at her without a response.

"What's she doing here anyway?" Shanna asked, wondering.

"It seems she needs a job and wants to work with us," Jack answered, his tone mocking.

"She's just your type, Jack," Celine teased. "As a matter of fact, I believe you and Steele have a little history between you."

"Oh, bite me, Celine," he uttered. She snarled playfully at him, baring her teeth.

"I'm glad you find this amusing, Celine," Will remarked sourly.

"This is really awkward for us," Jack complained.

"Why, because she threw your ass out of the Fleet?"

"Yeah, and the fact that what we are doing is illegal. Maybe it's a setup."

In contrast to Jack's irritation, Will was calm. "This is a bigger problem than we thought since the Fleet is the only line of defense for the human worlds against the aliens. If it crumbles, the worlds will fall, one by one."

"I hope this responsibility doesn't fall on us," Celine muttered.

"Then whose responsibility is it?" Will asked.

"It should be the Fleet's."

"You've got a point, but what can we do about it?" Jack asked.

"I think we'll need to step up our attacks on the supply centers," Will suggested. "That's one way I know of to slow down the aliens' advance."

"Bastille is close to completing the new cloaking device for the *Leviathan*," Celine informed them. "Once he's finished, we could take her out for a test run against the alien forces."

Will suddenly leaned against the table awkwardly. His face grew ashen. Shanna placed an arm around him for support. "I think you need to rest, my love." Will agreed and excused himself.

"Take care of him Shanna. He doesn't look well," Celine advised.

"I'm trying. It's not easy with Will." She escorted him out of the hall.

"How long has he been like this?" Celine asked Jack.

"Since Tenemon wounded him. Arasthmus thinks it's a residual effect from the Weevil poison, though."

"Can't he do anything for him?"

"No. Arasthmus gave him a full physical and couldn't find anything wrong. It's not unusual for him to have spells like these, two or three times a day."

"Something is definitely wrong with him."

"Maybe you should take it up with Arasthmus," Jack suggested. "He's the doctor."

"Maybe I will," she declared.

······●●●●●●●●●········

Shanna helped Will into his cabin on board the *Phantom* and sat him on the bed. She took his cane from him and set it in a corner. Will reached for her hand and squeezed it. He uttered weakly, "Thanks, Shanna. I don't know what I'd do without you."

"Are you going to be okay?" she asked, concerned.

"I'll be fine. I just need to rest for a bit."

Shanna couldn't sense his feelings or read his thoughts anymore. Since she had performed the Rite of the Dead on Will, they enjoyed a special telepathic bond, but something happened recently that ended that bond. "Will, you aren't telepathic anymore, are you?" she asked apprehensively.

"I'm not sure." He laid down on the bed and fell asleep.

Shanna nestled against him. She felt helpless to comfort him, not knowing what was wrong. More so, she fretted that he was blocking her from reading his thoughts, perhaps to hide something from her.

Will rested for several hours until he awoke, startled, with sweat beaded on his forehead. Shanna sat on the bed and watched worriedly.

"What's wrong, Will?" she asked.

"Nothing. I'm fine." he replied feebly,

Shanna looked at him suspiciously. His eyes were glazed and distant. "I'd feel better if I could share your thoughts and feelings again. At least then I'd know what's going on inside you."

"I would, too," he replied, frustrated. "I don't understand what's happening to me." She cradled him in her arms and they both drifted back to sleep.

Later, when they awoke, Shanna noticed that the color in Will's face was much better than before. Will grinned coyly until she asked defensively, "What's so funny?"

"You are."

"Why?"

"Because one little kiss from me means so much to you."

"Of course it does. You mean so much to me." She kissed him lovingly. Will felt renewed vigor and pulled her on top of him. "Now what?" asked Shanna.

"Now I know why I'm tired. You're wearing me out."

"Do you want me to stop?"

"Hell, no!" Will kissed her, but she pushed him away.

"Come on, Will. It's dinnertime."

"I'm not hungry."

"You have to eat. Maybe that's part of your problem." She took him by the hand and led him from the cabin.

When they entered the hall, everyone paused and looked up expectantly. They sat at the high table with Mariel, Breel and Laneia. Celine and Bastille sat together at one table in front of them. Steele sat at another table with Jack, Maya, Mynx and Neva. Steele stared at the Galactic Security Services girls and commented, "I know you, don't I?"

Mynx replied, "Yes, you do. What a small universe it is."

"Who are you?"

"I'm Mynx Falon. My partner is Neva Hirsch. We're former GSS agents who helped you with some of your problems a while back."

Steele shook their hands and remarked, "Fancy meeting you here. It seems Will has quite an assortment of friends working with him."

"We're all the best at what we do," Neva said, with pride in her voice.

"I realized that when I saw the *Leviathan* in the transport bay. What a coup that must have been."

"It was a masterpiece," Mynx responded, proudly like Neva.

"You say you are former agents for GSS?"

Neva replied bitterly, "Yeah. Those SOBs tried to kill us, and then they tried to stiff us on the reward money."

"I see they have their priorities," Steele kidded.

"That's okay. We taught them a valuable lesson. Besides, we're in a much better situation working with Will," Neva confessed.

"Boy, I'm jealous. Until recently, I always thought I had the best job in the universe as Imperial General of the Fleet. I'm really looking forward to working with this group of yours."

Servants brought in trays of food and ewers of wine. They poured glasses and set them before each person.

Will asked Maya curiously, "How did you get Steele away from the Fleet? I can't imagine they would let you just walk off with the Fleet commander without an explanation."

"I had to hide her inside a crate with our supplies."

Will turned to Steele and asked, "What do you think the Fleet will do when they discover you're missing?"

"I don't know. It would be an opportunity for them to replace me with a Weevil imposter, though."

"It will be more interesting to see if they don't replace you," he mentioned.

Maya looked at Will with heightened interest. "What do you mean by that?"

Will sipped from his glass and picked at his food. "Perhaps it works in their best interest not to replace her. If the Weevil are in control, then whomever they appoint will immediately arouse suspicion should you return."

"I see you think everything through," Steele remarked.

"The keys to success are tact and knowing your opponent," Will explained. "Since we don't always have the luxury of knowing our opponent, tact is very important."

Steele was anxious to learn more about Will and his band of renegades. She continued to probe with questions. "So, what's this I hear about a coronation and a wedding, Will?"

Will paused between bites and answered, "Shanna and I are to be married in four pogs. Immediately after the wedding, we're to be ordained King and Queen of Yord."

"Wow! How did that happen?" Steele asked, amazed.

Will pointed to Mariel. "This is Mariel, the head priestess on Yord. She can explain it much better than I can."

Mariel nodded to Steele and asked, "Have we met before?"

"I don't believe so, Mariel."

Will continued, "Mariel, this is Steele Furey. She is or was the Imperial General of the Space Fleet."

Mariel looked suspiciously at Steele and remarked coldly, "How interesting."

Steele extended her hand in friendship to Mariel and said, "It's nice to meet you."

Mariel shook Steele's hand delicately. She explained, "Will and Shanna are of royal blood. The women of Yord descend primarily from both their races, thus making the two of them the perfect candidates to rule Yord."

Steele was surprised. "I never realized Will was of royal blood."

"He's also Firenghian," Mariel mentioned.

"Wow!" Steele uttered, impressed.

Will asked her, "Did you know that Mariel was once a member of the Fleet?"

"No, I didn't," she responded. "Then what are you doing here, Mariel? The Fleet never lets good people get away."

Jack covered his mouth and faked a cough, saying, "Bullshit!"

Steele taunted, "I said good people, Jack."

Mariel explained, "I was banished here by the Fleet when they learned I was Firenghian. We are a race of shape-shifters you know?"

"I don't get it. Why would they do that to you?"

"They did that to any female that was different or deemed a threat."

"You didn't know about this?" Will asked, suspicious.

"No, I didn't. This is the first I've heard of such a practice."

Maya added, "If they found out I was Firenghian, I would have been banished here as well. I was lucky enough not to be identified."

"All of the women here were banished by the Fleet," Will pointed out.

"By whose orders?"

Mariel answered stoically, "Yours."

Steele was stunned and embarrassed. "I never gave those orders! I swear to you."

Jack sniped, "Obviously someone did."

"I guess I was blind to this. I never imagined the Fleet was so vulnerable to this kind of corruption."

"If the Weevil were behind this, then the Fleet has been compromised for quite some time," Will declared.

"There must be a reason the Weevil wanted these women banished," Maya remarked.

Will thought for a moment and then asked, "Has anyone ever seen a female Weevil?" Everyone pondered his question.

"All the Weevil at Eve's were male," Shanna mentioned.

"Nestor and his partner were males, too." Mynx added.

"Maybe I'll consult with Arasthmus," Will suggested. "There may be more to this than coincidence." He and Shanna stood and excused themselves.

Steele turned her attention to Celine, smiled and asked, "How can I forget you, Celine? You and your hair make a distinct impression."

"Why, thank you Imperial General," Celine answered pleasantly. She proudly pressed her hair down in the back.

"Please, call me Steele. I'm not the Imperial General anymore."

Maya introduced Breel and Laneia to Steele. "Breel and Laneia are Attradean," she pointed out. "They helped Will commandeer the *Phantom* and escape Attrades."

"I'm sure King Tenemon longs for the day he repays you for that ruse," Steele remarked.

"Laneia is Tenemon's daughter," Maya added.

Steele blurted out, "No way!"

"Yes I am, although I'm not proud of it," Laneia told her.

"How in the world did Will manage to get you, Tenemon's own daughter, to join him in the fight against him?" Steele asked.

"Will is a good and caring person. He freed Breel from my father's prison. We couldn't stay there so Will offered us a place with him. We've been friends ever since."

"This is truly extraordinary."

· · · · · · ●●●● ● ●●●● · · · · ·

Arasthmus and Regent played chess in the main quarters of the *Phantom*. Keira sat next to Regent and watched the match, curious over their intense rivalry. Will and Shanna entered the ship, interrupting them. "Arasthmus, I need your help."

"Not now, Will. This is a pivotal point of the match."

"It's important, Arasthmus."

"We can take a break, Arasthmus." Regent offered. "I don't mind."

Arasthmus sighed and tapped his fingers on the table. "Alright. What is it, Will?"

Will inquired, "What possible reason would the Weevil have for banishing women to Yord?"

"I didn't realize the Weevil were behind their banishment."

"Let's assume they were, hypothetically speaking. Why these women?"

"I'm not sure."

"We suspect this was done by Weevil posing as Fleet officers but what would they gain from such a thing?"

"Maybe they all had some degree of telepathic powers," Arasthmus suggested.

"Yes, but so did the males. Anything else?"

"Not that I can think of. Why?"

"The Weevil apparently have only mimicked men, so far as we know. All these women were banished here supposedly by the Fleet Commander's orders but she never gave those orders. Someone else did."

"So you think the Weevil purposely banished these women because they posed a threat to them?"

"Yes, I do."

"I'm not sure why a female's powers would be any different than a male's."

"Is it their powers or perhaps something else about them?" Regent asked.

"I'd bet on something else. Give me some time to think about it."

"Thanks, Arasthmus." Will stood up and studied the chessboard briefly. He remarked, "Knight takes rook, Regent."

"Get out of here, you brat!" Arasthmus bellowed as a friendly smile creased his face.

Shanna slapped Will's arm in dismay. "Only kidding. See you later," Will replied playfully. He and Shanna left the ship.

As they walked slowly across the dock to the stairs, Shanna chided, "That was mean, Will. You know they take their chess seriously."

"That's the fun of it. So, what do you think of all this? Any ideas?"

Shanna pondered and suggested, "What if there is something in the female chemistry that threatens the Weevil?"

"Like what?"

"I don't know. If we understood more about how they impersonate males, we might have a clue." They returned to the hall and sat at Maya's table.

"Well?" Maya asked,

"Nothing," Will told her. "Arasthmus is stumped, too."

Steele changed the topic and asked, "What's happening on the surface of the planet, Will?"

"Yord used to be a great economic power in the alliance," Will told her.

"I recall that very well. It was also a great time for peace and prosperity."

"But then the Attradeans put an end to that. That's why this operation we run underground has to remain a secret. Once we establish a government, we'll be able to enlist allies to protect us."

"Why do you need allies? Can't you protect yourself?"

"No, because the *Leviathan* and the *Phantom* are a secret. As far as everyone knows, we are rebuilding Yord for peaceful purposes and nothing more. Our alternative operation will enable us to build an economic metropolis on Yord by keeping our enemies at a disadvantage. Everyone must feel safe in coming here for Yord to thrive."

"Where did you learn to develop such bold strategies?" she asked, curious.

"At the Academy: American History on Earth – the Colonial Period."

"I may have to refresh myself," Steele confessed. "Obviously, I overlooked some very important lessons there."

Will held up his wine glass. "I propose a toast to our newest member, Steele Furey."

They raised their glasses and chanted, "To Steele."

Jack raised his glass halfway and coughed. Maya smacked him on the back of his head again. "Ouch!" he grumbled.

"Jack, behave or else," she warned him.

"Or else what?" he asked, wondering how far she'd go to defend Steele. She whispered something in his ear and Jack frowned. He knew he lost this battle.

"Thank you, everyone. I'm grateful to be here with you," Steele told them. She asked Will, "What do you think about forming a new alliance to replace the Fleet? You seem to have the right plans to build a competent union."

Will sipped from his glass and replied, "Why a new alliance?"

"I don't think we can salvage what we have."

"Perhaps, but it's too early to make that decision," he explained.

"And maybe not," Steele conceded.

"I can't help feeling apprehensive about the Boromians," Will remarked. "The Weevil work closely with Tenemon and the Attradeans for now. If the Attradeans were taken out, what would the Boromians do? I think they're letting the Attradeans fight their battles. Whoever wins will be in a weakened state. The Boromians could come in and clean up the leftovers and take over the alien alliance."

"I don't put it past them," confessed Steele. She seemed concerned by Will's perception of the war.

"I also think we need to take a crack at the Fleet's upper echelon," Will mentioned. "We've got to know how far the Weevil penetration has gone."

"What good would that do?" she asked.

"We'll need to salvage whatever resources we can. Even if the Fleet fails, there are a lot of ships and crewmen out there. We need them on our side."

"And how would you do that without tipping off the Weevil?" she inquired, interested in a potential plan for this.

"There's got to be a way," he replied confidently. We just haven't thought of it." Suddenly, he became light-headed and tilted helplessly against Shanna's shoulder.

She put an arm around him. "It's time for Will to rest. He's recovering from a serious injury."

"Why don't we all get a good night's rest? It sounds like we have a big day tomorrow," Maya suggested.

Shanna bade them goodnight. "I'll help you with Will," Mariel offered.

Will breathed heavily and broke into a cold sweat. He uttered weakly, "I'll be alright. It's just a little warm in here."

"Just be quiet and let us help you," Shanna scolded him. She and Mariel escorted Will back to the *Phantom* and into bed.

"Are you sure you're alright, Will?" Mariel asked.

"I don't know. One minute I'm fine and the next I feel all messed up." The women laid him on his back and removed his boots. "It only happens a few times a day," he added.

"That's a few times too many and it's getting worse," Shanna complained.

"What do you think of Steele, Mariel?" Will asked, his voice weak.

"I don't trust her."

"Why not?" Shanna asked. "She seems okay."

"There's something about her that's not right." Mariel complained.

"Do you think she's Weevil?" Will asked.

"No. But I think she's something else. I just can't put my finger on it."

"We could keep an eye on her in case she's up to something," Shanna suggested.

Will placed his hands over his face and uttered sadly, "Oh this is getting old."

Mariel asked Shanna, "How come you don't get these spells?"

"Should I?"

"If it's a result of the Weevil poisoning, you should get them too."

"But I feel fine."

"And that's what makes it odd."

"I'll be okay, Mariel. Just give me time," Will assured her.

Mariel's look was grave. "Why don't I believe you?" she replied and left the cabin.

Shanna undressed Will and covered him. He fell asleep quickly. She undressed and lay next to him.

· · · · · · · · ● · · · · · · · · · ·

The next morning, everyone met at the hall for breakfast. Steele entered, wearing tight blue jeans and a white blouse. She wore her hair down upon her shoulders. Will and Jack couldn't help but notice how attractive she was out of uniform.

Shanna poked Will in the ribs and chided, "Hey, I'm over here."

"I'm sorry. I can't believe that's the same person who wore a uniform and a scowl for all those years."

"Yeah, there really was a woman under all that misery," Jack agreed. Maya frowned at him, annoyed by his negative remarks.

Jack kissed her cheek. "Don't be jealous, Maya. I still love you."

"I certainly hope so. If not, I'll take you back to your space station, if there's anything left of it." Will and Shanna burst into laughter and Shanna high-fived Maya.

Jack complained to Will, "Thanks for the support, buddy."

"Sorry, Jack."

Steele sat at their table and announced pleasantly, "Good morning, everyone."

"Good morning, Steele," Maya greeted her.

"I see everyone is having a good time."

"Hardly," Jack quipped.

Steele hugged him and said, "Come on, Jack. I thought you were tougher than that." He grimaced. Then she inquired, "What's the game plan for this morning?"

"Maya and Saphoro will cover us in the *Luna C* in 'cloak' mode. Celine and Bastille will pilot the *Phantom* and handle forward firing control. Jack and I will work the turrets until we're on the ground. Then, we'll enter the supply depot. Shanna, Breel and Laneia will assist me on the ground."

"And what about me?"

"Do you want to go on the *Luna C* and monitor our operation?" Will asked.

"I'd much prefer to be in on the action," Steele responded, leaning forward in an intimidating pose.

"Then you'll go with us on the *Phantom*. Jack will make sure you have a weapon when we hit the ground."

"What about the *Leviathan*?" she asked, pressing the issue once more.

"No *Leviathan*. It's not supposed to exist, remember."

"Ah, that's right. You made a hefty sum of credits from the Fleet for destroying her." Will became suspicious about her concern to bring out the *Leviathan* and wondered if she did have an ulterior motive for being there.

Jack mocked her, "Capitalism is a beautiful thing."

"Jack will open a non-alliance account for you. Whatever we make, we split up among the team," Will informed her.

"So that's why my officers couldn't find you. They watched every alliance account, particularly since we suspected a member of the Fleet was behind the job."

Will patted Jack on the back and responded, "Jack did a masterful job of arranging things."

Sounding bitter, Jack interrupted, "Let's go, everyone. It's show time." Will was amused at Jack's behavior as they departed the hall and boarded the two ships.

Will gave Steele a tour of the *Phantom* as they flew toward their target. She was particularly interested in the cannons. She climbed into the turret and felt the controls. Will asked from the bottom of the ladder, "Have you ever fired before?"

"Yes, I did. I flew on the *Luna C*'s last combat mission before she was refitted for clandestine surveillance. I shoot very well."

"Perhaps you'll have a chance to show us."

As Steele climbed down from the turret she declared, "Let me make one thing clear, Will. I don't like to lose. What's happened to the Fleet under my watch is hard for me to swallow. I will fight to the end if that's what it takes to exact my revenge on the Weevil."

"Just remember, Steele, the key to our success has been tact. We have to be smart to be successful. I hate the Boromians more than anything, especially since they killed my parents, but I have to pick my battles with them. You will, too."

"Were you aware that, if Maya hadn't requested to pick you up from the academy, you would have been killed, too?"

Suspicious, Will questioned Steele, "Why wasn't anyone aware of an armed hovercraft on the Fleet Academy's own grounds?"

"One of my officers was responsible for security, especially where the *Luna C* and the *Ruined Stone* were concerned. I didn't think much of it, although he was very persistent about handling it. Maya convinced me her mission was critical and that it was more important for her to meet with you secretly at the graduation. I never imagined your parents would be killed in route and assassins would come after you as well."

"So you think it was a conspiracy to kill me and my parents?"

"It could have been. Is there a reason the Weevil would want them dead, too?"

"Not that I know. I did learn that the Fleet delayed their search for them. My parents held out for quite some time before the Boromians got to them."

"I'm sorry. I feel responsible for everything that's happened under my command."

"Well, all we can do now is make them pay."

"And that they will."

Bastille announced over the page, "We'll be landing soon. Man the turrets and happy hunting, friends."

Will climbed the short ladder into the turret. As he leaned into his seat, he became light-headed and tumbled to the floor, unconscious. Steele heard the thud and shouted, "Will, are you okay?" There was no answer.

"Will!" Steele shouted again. She hurried down the ladder from her turret and found Will lying unconscious on the floor. She quickly sat him up and called for help.

Shanna and Laneia rushed into the bay. Shanna saw Will and was devastated. "What happened to him?" she asked.

"He was in the turret when I left him. Then he collapsed and fell to the floor."

Shanna hugged Will and said soothingly, "I'm here for you, Will. Please be alright."

Steele instructed the girls, "I'll take the turret. You take care of Will."

Shanna and Laneia lifted Will to his feet and helped him to his cabin. Shanna felt his forehead and fretted that he burned up with fever.

"Shanna, why don't you stay with him?" Laneia suggested. "We'll take care of the depot."

"Thank you, Laneia." Laneia left them and returned to the main quarters.

Jack handed pulse pistols to everyone. He noticed Will and Shanna were missing. "Hey, can someone beckon the lovebirds?" he kidded.

"Will is hurting again. Shanna's staying with him," Laneia explained.

"Where's Steele?"

"She's in the turret."

"That's just great – a rookie in the turret. Get ready for a fun landing."

Jack hurried to the turrets. He stopped at the ladder to Steele's turret and called to her, "You know what you're doing up there?"

"Just watch me."

Jack reached up the ladder and handed her a pulse pistol. "You'll need this when we land."

"Thanks, Jack."

He noticed she smiled at him for once. "You're welcome. Good luck."

For a moment, he considered that maybe there was a side to her he could learn to like. But then he recalled their past and thought cynically to himself, "Nah, that's impossible." He climbed the ladder to the other turret and strapped himself in.

The *Phantom* glided toward the Ceratoan depot, just above ground level to avoid detection. At such close proximity to the ground, the automated firing system was inoperative and all firing had to be done manually. Jack and Steele spotted several armed vehicles. Jack fired three shots and took out the lead one. Steele fired two shots and took out the next two. Jack fired once into the driver's cab of a personnel carrier and it burst into flames. Steele fired at the next carrier, which also burst into flames. A friendly rivalry developed between them as they targeted each vehicle. Jack was impressed with Steele's accuracy on the cannons, particularly at long-range. He wondered how much previous experience she could have had on the cannons with her rank.

As they drew closer, they spotted anti-aircraft cannons near the depot. The cannons couldn't target them accurately so long as the ship stayed close to the ground. Jack and Steele peppered the guns with cannon fire and destroyed all six with ease.

The *Phantom* landed a short distance from their target - a long, cinderblock warehouse. As soon as the hatch opened, Breel rushed out and took cover behind one of several crates. Three Ceratoan soldiers rushed toward the ship, but Breel was ready and fired at them. Two fell to the ground, dead, while the third ducked behind drums of chemicals and returned fire. Breel crept across the grounds toward a small tool shack near the warehouse entrance. The soldier fired several more shots at him but missed.

Laneia raced out of the hatch and disappeared among the crates. She crept up on the soldier from behind and snapped his neck. With a wave of her clawed hand, she signaled to Jack at the hatch that it was clear.

Jack and Steele rushed to the warehouse doors and joined Breel. Jack pushed the hinged, aluminum doors open and scanned the area inside. There was no one in sight. Jack and Steele slipped inside and parted in opposite directions, leaving Breel to guard the doors.

A dozen soldiers appeared from a side entrance and spread out behind crates and forklifts. They pinned Jack and Steele down with a steady barrage of pulse fire. Breel slinked toward the corner of the warehouse and gestured for Laneia to join him. They climbed a ladder to the roof and perched over a large window. Breel lifted the window and latched it open. He and Laneia then targeted the unsuspecting soldiers below and picked them off, one by one, until their pistols required charging time.

Three more soldiers rushed across the open floor of the warehouse, but Steele killed all three with abnormally good marksmanship. For the moment, the coast was clear. Jack was impressed by her shooting and met her in the middle of the floor. "Damn, you're deadly Steele."

"Years of practice and lousy men will do that to a woman," she kidded.

"What else does it do?" he taunted.

"Come on, Jack. Was I that tough on you?"

"Why do you think I left?"

"I booted you out, remember? You didn't volunteer to leave."

"Yeah, yeah, yeah. Can you drive a forklift?"

Steele asked disappointedly, "That's it? End of argument?"

"Maybe we'll pick it up at another time."

She sighed and relented. "Fine. I'll drive the forklift."

"I'll point out the crates we want. Drop them inside the hatch so Bastille, Breel and Laneia can lower them into the cargo bay."

Steele manned the forklift and followed Jack. He broke open several crates of bottled water and motioned for Steele to take them. Steele picked up the crates one by one and delivered them inside the hatch.

Laneia and Breel scurried down the ladder from the roof to the ground. They rushed to the ship and stepped inside the hatch. Breel operated the lift and lowered the crates to the cargo bay below.

Jack pointed to two more crates. As Steele picked up the first one, she spotted a soldier creeping behind Jack. She backed the forklift slowly and aimed her pulse pistol at the soldier's head. As the soldier targeted Jack,

Steele fired first, striking the unsuspecting soldier between the eyes. He stood still for a few seconds, as if stunned, and then fell to the ground dead.

Steel's shot had passed precariously close to Jack's head. He felt the searing heat from the shot on his ear and was startled. When he saw the soldier on the ground behind him, he glared at Steele. She smiled, gave him a thumbs-up, and drove the forklift with another crate to the ship.

Jack found two large crates with vaults in them. He broke the crates apart and examined the vaults. Steele returned with the forklift and awaited instructions. Jack directed her to put the vaults inside the hatch, too.

"What are they?" she asked.

"I don't know, but they look official enough."

Steele took the vaults to the ship with Jack walking behind, searching for soldiers.

· · · · · · ● · · · · · · · ·

Maya and her co-pilot, Saphoro, watched the monitor anxiously as a convoy of assault vehicles approached the depot. "Looks like we've got company, Saph," Maya indicated.

Saphoro replied, "I'll take care of them." She programmed four torpedoes and pressed a button on the armament control panel. Four red energy pulses darted from the *Luna C* and struck several of the vehicles, sending up multiple plumes of smoke and bright explosive bursts. "That should slow them down some," Saphoro commented proudly.

"Nice work, Saph."

"I've been practicing on the weapons systems."

"How so?" Maya asked.

"Oh, Bastille made a program, which works like a simulator."

"No kidding."

"He said they use them on Earth. They're like video games."

"That man is full of surprises."

· · · · · · ● · · · · · · · ·

When the forklift approached the hatch, Breel guided Steele as she set the vaults just far enough inside the hatch so they could close it. Breel lowered the platform to the deck below where the cargo would be moved and then stored by automated equipment. Fireworks lit up the sky from a short distance away, catching their attention.

Let's step it up, Jack," Steele advised. "Looks like we've got company nearby."

"Nag. Nag. Nag," he teased. Jack carried a knapsack across the warehouse to a row of crates out of site from the hatch. He took out a package of explosives and set the electronic detonating device. Suddenly, he found himself on the ground, pummeled from behind.

Steele helped Laneia and Breel load the last supplies and boarded the ship. Jack hadn't returned so she hurried back inside the warehouse to search for him.

When she entered, two Ceratoan soldiers were beating Jack. Steele tapped the first one on the shoulder. When he turned around, she blasted his face off with a close-range shot from her pulse pistol. The second one turned around, poised to fire, but Steele kicked the pistol from his leathery hand and stuck the barrel of her pistol in his fanged mouth. She asked cynically, "Any last words?" The soldier mumbled something incoherent. Steele laughed at him and said, "Sorry, pal. It's not your day." She pulled the trigger and blew out the back of the soldier's head and neck. He fell to the ground, lifeless.

Steele helped Jack to his feet. His face was swollen and his lips were bloody. His eyebrow was cut and the stitches in his chin burst open. "Are you okay?" she asked sympathetically.

Jack straightened his shirt and muttered, "Sure. I had everything under control."

"Of course you did."

"But thank you, just the same. I may have been at a slight disadvantage."

Steele picked up the electronic detonating device and programmed the timer for three minutes. She set the package between two crates and walked away.

"Cutting it a little close, aren't we?" Jack asked.

"You said you were fine. Show me."

"Women. Why me?" he muttered. They hustled back to the ship. Once inside, Steele closed the hatch and sealed it. Jack shouted into the page, "We're in, Celine. Get us out of here!"

"Roger, Jack."

The depot exploded in a fiery mass. When the ship rocked violently, Jack fell on his ass, much to Steele's amusement. She gripped the handles on a cabinet tightly for support. The area surrounding the depot burst into flames as well. Intense pulse fire from ground batteries rocked the *Phantom* as it raced away.

Celine's voice sounded sarcastically from the page, "Nice, Jack! Blow us up next time."

Jack glared at Steele again and answered humbly, "Sorry, Celine. We cut that one a little close."

"A little? The engine temperature jumped seven degrees from that blast."

Jack remarked sullenly, "Thanks a lot, Steele. I'm sure I'll get all the flak for this."

"You were too slow. Next time, run faster." She offered him a hand and helped him to his feet. Jack patted her on the back.

"Nice work. Oh, and thanks for saving my ass earlier, although it took you long enough."

"My pleasure. Even for you." They laughed together.

"I'm going to check on Will. I'll be back in a while," Jack said.

"I'll help your friends with the supplies."

Jack hobbled down the hall and entered Will's cabin. When he arrived, Shanna was in tears. Will's face was ashen and he was soaked in sweat. Jack asked nervously, "What happened, Shanna?"

"He was convulsing. I couldn't stop him, Jack, and now he's unconscious."

Jack examined Will and felt his pulse. "His heart is still beating."

"No kidding!" Shanna yelled. "That doesn't help."

"Sorry. I'm not a doctor." Jack listened to Will's breathing. It was steady and strong. "It seems like he's okay now. I'll tell Celine to get us back to Yord ASAP."

Shanna sobbed and hugged Will. "I'm really worried about him, Jack."

"Believe me, Shanna, I am, too."

Jack was shaken at the sight of Will lying helpless. He somberly returned to the cargo area. Steele knelt in front of the first vault and studied the locking device. "Any ideas?" Jack asked.

"This looks similar to a device we use to protect secret information. If you tamper with the locking device, the contents of the vault will be destroyed."

"So you think these vaults hold important information?"

"I'm sure of it. Vaults like these are very expensive. Why don't you take a break? You took quite a beating back there."

"I've had worse."

"Oh, Jack. You don't need to play the strong warrior role with me," she kidded.

"That's it!" he groaned. "I'm leaving." Jack limped to the other side of the cargo bay and inspected the crates and cases they took from the depot. Breel and Laneia stacked some of the new supplies in cabinets and closets. Jack quipped, "Not bad. Huh, kids?"

"Nice haul, Jack. You and Steele did good." Breel complimented him.

"Gee, I never thought I'd hear that from anyone."

· · · · · · ●●●●● ● ●●●●● · · · · ·

The *Phantom* returned to the base on Yord and glided into its berth. Jack hurried up to the temple and sought out Arasthmus.

Mariel sat with Arasthmus at a table in a small room at the end of the corridor. They discussed medicine and healing arts. Jack rushed in and interrupted, "Will needs help, Arasthmus. He's unconscious."

"What happened?"

"He blacked out. Shanna said he was having convulsions."

"Damn that boy! I'll be down to check on him." He quickly left the atrium.

"Could Will have eaten anything that might cause these lingering problems?" Mariel asked Jack.

"He and I eat about the same."

"There's got to be an explanation for this."

"I wonder if the Weevil poison really does have something to do with this," suggested Jack. "Maybe he isn't fully cured."

Arasthmus returned with a leather bag in his hand. "What's this about the Weevil poison?"

"Are we sure it was poison the Weevil injected into Will?" Mariel asked.

"What else could it be?"

"You're the doctor. You tell us," Jack retorted.

"Maybe we'll need to approach this from a different direction," Arasthmus said.

Jack suggested, "Maybe this could be the key to why the Weevil are attracted to men only."

"Don't jump to any conclusions until we understand what's going on with Will's body," Arasthmus cautioned. They proceeded to the *Phantom*.

II

CONSPIRACIES AND COUNTERMEASURES

Arasthmus, Mariel and Jack barged into Will's cabin. Arasthmus opened his bag and removed a syringe and a vial. After filling the syringe, he injected the contents into Will's arm. "What's that for?" Shanna asked.

"To stabilize his metabolism." Arasthmus felt Will's pulse. He thought for a moment and then questioned Shanna. "How many close encounters have you had with the Weevil?"

"At least three."

Arasthmus took another syringe from his bag and extracted a blood sample from Will's arm. "Did any of those creatures bare claws in the prior encounters like the ones we took out of Will's shoulder?"

"No." She thought about it and then added, "That's strange. Why only Will and why only at Eve's?"

"What's so strange about it?" Jack asked.

"When we were on Nestor's ship, neither he nor the other Weevil bared claws at us either," Shanna mentioned. Jack grew more confused and waited for her to explain herself. "If the claws were a weapon, why did they only use them at Eve's and why only on Will?" Shanna pondered aloud.

Arasthmus asked, "Does anyone know how the Weevil impersonate humans?"

"We assumed they replicated their targets," Jack answered.

Arasthmus turned to Mariel. "Find Mynx and bring her here. She's probably on board the *Leviathan*." He then pressed several areas on Will's body with his fingers.

Mariel suggested, "What if they don't molt but actually use the body as a host?"

"How could they do that?" Shanna asked, perplexed.

"Mariel might have had a valid point," Arasthmus commented. "Maybe Will wasn't poisoned. Maybe this is the means they use for taking over a host." He furrowed his brow in thought.

"You mean Will's becoming a Weevil?" Shanna cried out.

"Maybe the convulsions are a result of his body fighting it," he suggested. "Will's not like other men."

"He told me that male Firenghians can't pass on the traits that made his father a shape-shifter," Shanna revealed. "Only women can."

"Well, how did he pass them on to you?" Jack asked.

"Um, well, I was already a shape-shifter like Will, just from a different race. I picked up some of his telepathic abilities when I performed the Rite of the Dead on him."

"Does he know that?"

"I, uh, I'm not sure," she muttered, realizing she had secrets from Will that were never discussed.

Arasthmus retrieved another syringe from his bag. "Hold out your arm Shanna," he instructed.

Shanna gasped. "What's that for?" She winced briefly from the prick of the needle.

Jack watched curiously while Arasthmus drew a sample of blood from her arm. Arasthmus then took a small electronic instrument from his bag. He pointed it at Will and scanned his shoulder area. The instrument beeped several times. The instrument beeped again when he scanned Will's abdomen.

Jack and Shanna became edgy. "What's that mean?" Jack asked.

"It means I'm going to operate on Will right now."

"Now? What for?" Shanna asked, while placing a hand on Will's shoulder protectively.

While putting on plastic gloves, Arasthmus explained, "There are two foreign bodies inside him and I think they need to be removed immediately." He tied Will's wrists and ankles to the bed, and then removed a small kit from his bag and opened it.

"Shanna, perhaps you should wait outside," he suggested.

"No. I'm staying," she said resolutely.

"Then brace yourself. I don't know what we're going to find."

Arasthmus unbuttoned Will's shirt, showing his chest and abdomen. He sprayed Will's abdomen with a cold disinfectant.

Jack observed Arasthmus' actions closely. "What do you think you'll find, Arasthmus?"

"I don't know and that concerns me greatly."

Jack and Shanna were shaken by Arasthmus' fear. He never exhibited much emotion and to hear his concern frightened them.

Arasthmus cut a four-inch incision in Will's side then another incision perpendicular to the first. He parted the flesh and scanned Will's cavity with the electronic device. The instrument beeped steadily. He reached into the opening with a forceps and tugged gently until an oval-shaped object appeared at the incision. Numerous tendrils stemmed from the object. "This could take a while and I don't know if I can do this," admitted Arasthmus, unsteadily. "This is one complicated mess."

"Is there anything we can do?" Jack offered.

"Pray and I mean that in the utmost manner." Jack and Shanna looked to each other for support but neither was confident.

After three hours of maneuvering the tiny tendrils and snipping them, Arasthmus finally removed a four-inch pod with thin, dangling tentacles. He held the instrument near the pod and it beeped rapidly.

Jack and Shanna stared in horror. Jack's mouth opened, closed, and then opened again. "What the hell is that thing?"

Arasthmus used his free hand to dump flowers out of a nearby urn. He dropped the pod inside. Shanna looked nauseous. "What was it, Arasthmus?"

"A Weevil seed perhaps."

She gagged and coughed. "Oh, no! And there's still another?"

"I'm afraid so." Arasthmus stitched Will's abdomen closed and then sprayed it again with the disinfectant.

Tears streamed down Shanna's cheeks. Jack tried to comfort her but she pushed him away, determined to be strong in front of the men.

Arasthmus repositioned Will and sliced into the flesh under Will's front shoulder area. He reached in with the forceps and pulled another pod near the surface of the incision. He broke into a sweat and wiped his brow with his sleeve. Two and a half hours later, he snipped the last tendril and removed a three-inch pod. He dropped it into the urn with the first one. He then stitched the incision and sprayed it with disinfectant.

"Is that it?" Jack asked.

Arasthmus scanned Will from head to toe but received no further beeps. "I believe so," he uttered, relieved to be finished. He flopped into a chair and wiped beads of sweat from his forehead and cheeks.

"Thank you, Arasthmus," Shanna said appreciatively.

"Don't thank me yet, young lady. Let's wait until he recovers."

Shanna squeezed Will's hand and rested her head against it while she silently prayed.

Mariel, Mynx and Neva entered the cabin as Arasthmus removed his bloodstained gloves and stowed them in a plastic bag. They were surprised to see Will's condition. "What's going on Arasthmus?" Mynx asked.

"You mentioned before that you never saw anyone survive one Weevil poisoning, let alone two."

"Yes I did. Why?"

"Were you sure the infected individuals were dead?"

"Well, I assumed so. I never followed up on the bodies after they were taken away."

"So the victims may not have been dead."

"It's possible, but unlikely."

Arasthmus handed the urn to her and instructed her to take a look. Mynx inspected the pods inside the urn and looked disgusted. "What the hell are these?" she asked.

"They were inside Will."

Neva looked inside the urn and exclaimed, "Holy shit!"

Mynx apologized emphatically for her oversight. Arasthmus patted her on the back and said, "Don't worry about it. We learned something very valuable today that could affect the outcome of the war."

Shanna raised her head and exclaimed, "Hey, I can sense Will's feelings!"

"Are you sure?" inquired Arasthmus.

"Oh, yes. He's relieved and pain free."

"Keep him in bed until the incisions heal. I have to examine these pods."

"I certainly will!" Shanna wiped her tears away and rested her head on Will's bare chest.

"Will he recover?" Mariel questioned Arasthmus.

"I'm sure he'll be his old self before long."

Shanna hugged him and blurted, "Thank you so much, Arasthmus."

"You're welcome, dear." Arasthmus eagerly left the cabin with his bag and the urn.

Jack announced edgily, "I've got to take care of some things. I'll check back later."

Mynx was trembling. "I can't believe I didn't see this happening."

Neva patted her on a shoulder. "Remember when we killed Nestor and the flesh fell off of him?"

"Yeah."

"Imagine if you folded up the legs on that creature and pushed in the face; it would look just like one of those pods."

Mynx thought for a second and answered emphatically, "You know, I think you're right, Neva."

Shanna sensed Will's consciousness returning and shouted, "He's back! He's going to be okay." Will opened his eyes and placed his hand on Shanna's cheek. Relieved, Mynx and Neva smiled at him.

"What happened?" Will asked, baffled.

Shanna kissed him gently. "You're all better."

Will felt the stitches on his abdomen and shoulder and asked nervously, "What's this all about?"

"Arasthmus removed two foreign bodies from inside of you."

"Foreign bodies?" he asked, surprised. "What kind of foreign bodies?"

"Weevil pods. You might have given us the answer to the Weevil."

Will look perplexed. "How did that happen?"

"It doesn't matter right now."

Maya and Neva slipped out of the room, allowing Shanna and Will some privacy. Mariel winked at Shanna and stepped out as well.

Shanna cried, "Oh, Will, I'm so glad you're back!" She hugged him tenderly. Tears washed down her face.

"But I never left."

"Yes, you did. When we lost our bond, I felt like I lost an important part of you." She kissed him passionately then latched onto him and tightened her arms protectively around him.

Will teased, "Easy, now. I'm not going anywhere."

"You bet you aren't. I want to make sure you are a hundred percent for the wedding," Shanna declared. Will laughed at her and held his hand against her cheek.

· · · · · · ●●● ● ●●● · · · · · ·

Arasthmus entered his lab with the urn and dumped the pods into a transparent container. He removed the vial with the sample of Shanna's blood from his bag and set it on the table. Mynx and Neva entered, eager to see what he was up to. "You're just in time," Arasthmus said.

Mynx commented, "For something good, I hope."

"I think *interesting* is a better word." He took a small syringe from a drawer and drew some of Shanna's blood from the glass vial. He injected one of the pods with the contents and watched.

The pod shook violently and exploded. Dark green and brown pieces of gelatinous material ran down the side of the container.

Neva cried, "Oh, that's so gross!"

Curious, Mynx asked, "What do you think is going to happen when you inject Will's blood into the pod."

"Let's find out." Arasthmus retrieved Will's sample from his bag and set the vial on the table. He took another small syringe from the drawer and extracted some of the sample.

Neva asked, "Is this going to be gross, too?"

"What do you think?" he countered. The women looked unsure of themselves.

Arasthmus injected the remaining pod with Will's sample. The pod throbbed for several seconds and returned to dormancy.

"That's it?" Mynx questioned, wondering why.

"Yes it is! This is astounding!"

Neva asked, "So what does it mean?"

"It means that now we know how the Weevil are taking human hosts. Plus, we know that something in the traits of the female blood types destroys them. All we have to do is figure out how to use it against them."

Mynx asked, "What do you propose to do next?"

"Lots of testing. I stopped practicing medicine a long time ago to be a politician. Now, it looks as though I'm back to practicing medicine full-time again."

"I'll inform the others about your results, if you like," Mynx offered. She and Neva left Arasthmus alone in the lab.

They met Mariel outside Will's cabin. Mariel sensed their eagerness and asked, "Did something happen? You two look like you're in a rush."

"We've got some very interesting news on the Weevil pods," Mynx announced.

"I see," mentioned Mariel somberly. "Unfortunately, I think there are more pressing issues that may require the attention of your group."

"Today hasn't been a very good one so far. What is it?" Neva asked.

"I need to see Will and Shanna immediately. Come with me." They followed Mariel into Will's cabin.

Shanna sat next to Will's bed and held his hand, while he giggled at something she said. They looked up, surprised to have visitors. Mariel inquired, "How are you feeling?"

"Much better. I'm fortunate Arasthmus is so adept with his medical skills."

Mynx announced, "We have good news about the pods Arasthmus took out of you?"

"Two of them, huh? What a surprise?"

"Besides that. We think those pods prove that the women were banished to Yord by the Weevil for a reason."

Shanna asked, growing anxious, "What did you find out?"

"Arasthmus injected one of the pods with a sample of your blood and it exploded."

"My blood?" Shanna replied, disbelieving.

"What did it explode from?" Will asked.

Mynx held up a finger for patience and said, "It gets better. When he injected the other pod with your blood it had little effect."

Will looked more perplexed. "I don't understand."

Neva related in simple terms, "First of all, the Weevil didn't poison you. They tried to take over your body. The pods apparently grow inside you until they become you."

"No kidding!"

"Your body was trying to fight the Weevil pods."

Shanna pondered aloud, "Maybe that's why your telepathic capabilities stopped."

"Well, they're back now and I feel like new. But why did Shanna's blood destroy the pod?"

Neva pointed out to Will, "You told Shanna that only the women can pass on traits in your races. Well, something in those traits can destroy the Weevil. If the men can't pass along traits, they can't affect the Weevil pods." Will was astonished by the revelation.

Mariel asked Shanna, "Have you sensed anything from the Eye?"

"No, I haven't. I've been focused on Will's condition."

"I felt something from the Eye but my perception isn't completely restored yet," Will commented.

"A great dragon will succumb to another great dragon through deceit. Do you know what that means?" Mariel asked. Shanna focused hard to learn the revelation.

Will considered her words for a moment. "Sounds like Tenemon. Perhaps he's going to turn on the Boromians."

"Is that a good thing?" Neva inquired.

"Probably not. I've got to get up."

"Will, don't you dare!" Shanna admonished him.

"I'll be careful. I can't stay here like this."

"You are so damn stubborn, Saris!" Shanna blurted angrily. Mariel took Shanna by the hand and led her to the door.

Shanna resisted slightly. "What are you doing, Mariel?"

"We're putting Will under house arrest." Shanna grinned at the idea and the women left the room.

Will called out frantically, "Wait! Don't do this to me."

Mariel replied as she closed the door, "It's for your own good."

"I can't let women run my life like this!" he exclaimed and then resigned himself to his fate.

• • • • • • • • • ● • • • • • • • • • •

A few weeks passed before Will recovered from the surgeries. Shanna, Mynx, Neva and Mariel entered his room and stood by his bed. Will taunted them, "Well, well, well. The executioners have arrived."

Mynx teased him, "Keep it up and you'll spend another three weeks in here."

"So what, pray tell, do you ladies want of me?"

"We think it's time to do something about Tenemon," Shanna declared.

Will responded sarcastically, "Ah, now that we've given him a head start on us."

"We've monitored the Eye and he's done nothing until now that couldn't wait," Mariel informed him.

"I knew that. I can read the Eye too, you know. Now get me out of here."

Shanna reluctantly helped him sit up. "Remember, you jeopardized the last mission because you weren't one-hundred percent,"

Mynx also scolded him, "You'd better not bust your stitches open or else."

"I get the point, already. Now, help me get to the hall. We'll assemble everyone and prepare our next move." Neva volunteered to round them up.

"That's right. Leave us with the dirty work," Shanna teased. Neva and Mynx giggled over her comment.

"Hey," Will hollered. "Since when did I become dirty work?"

Mynx poked him gently and warned, "Shut up or we'll leave you here by yourself - again." Mynx and Shanna lifted Will to his feet. They carefully escorted him from the *Luna C* to the hall at the upper level of the base.

Will pulled his arms away from them and grumbled, "I can walk on my own. I'm just a little sore."

They looked doubtfully at him but gave him space, allowing him to hobble inside the hall. He and Shanna sat together at the first table instead of the head table. Mynx took a seat across from them and commented, "I'm really glad you're alright. I was worried about you."

"Thanks, Mynx. I finally feel like I'm returning to normal."

"You must have fate on your side," she remarked. "To survive what the Weevil did to you is truly unbelievable."

Mariel, Jack and Steele entered the hall. Jack extended his arms out to Will and joked, "Well, buddy, are you going to live or not?"

Will was embarrassed and replied humbly, "I think so."

Mariel, Jack and Steele sat at Will's table. "Where's Maya?" Will asked.

"She took Breel and Laneia with her to scout the Attradeans. They'll be back tomorrow," Jack explained.

Neva returned to the hall with Celine and Bastille. She informed Will, "This is everyone, except for Arasthmus, Regent and Keira."

"Let's get started then." Will stood up uncomfortably with help from Shanna and announced, "We suspect that Tenemon is preparing to coerce another alien race to join his fold. We have to stop him."

"And how do you propose to do that?" Celine asked.

"As soon as we find out who his target is, we'll pay them a visit. Perhaps we can set a trap for him."

Steele looked doubtful. "That sounds all too easy. What if the targeted race isn't open to discussions with us?"

"Then we'll use Plan B."

"Which is?"

"I'll let you know as soon as I know. We'll wait until we hear back from Maya before we make our move."

Jack asked uneasily, "You aren't planning to participate in this mission, are you?"

"Why wouldn't I?"

"Last time, we had to scrape you up off the deck."

Will assured him, "I'll be fine now."

"We're counting on you. Don't screw up."

"Yes, Dad." The women chuckled at his response. Will continued, "So we now understand how the Weevil take hosts."

Mynx joined in. "Yes, and it's quite gruesome."

"How sure are you of this?" Jack asked, staring at Mynx.

"Oh, we're quite sure. I witnessed the test."

Shanna added. "And now we believe we understand why women were banished here on Yord: they were a threat to the Weevil. We also understand how they take over a host. They insert tiny pods into the host, which grow within until they physically become the host."

"Arasthmus is looking for a way to apply what we've learned as a weapon against the Weevil," Mynx announced.

Steele asked Will, "Does this mean that we have a shot at saving the Fleet?"

"Quite possibly. Tell me, Steele, do you know the commanders on the Fleet vessels well enough to distinguish them from Weevil?"

"I think so. I made it a point to know my commanders very well before I sent them out."

"Mynx and Neva will take you on board the *Leviathan*. See if you can contact any of them using its discreet communication system."

"Then what?"

"Tell them to be ready for instructions. We're taking back control of the Fleet. That's all for now." Everyone departed the hall except for Shanna, Will and Mariel.

"Guests are already arriving on the surface for the wedding," Mariel reminded them.

Will complained, "They're a bit early, aren't they?"

"I believe they're evaluating our strength to function as an economic power and a ruling entity. I know that three engineers from Tarsus examined the new palace buildings as they were being built. They asked many questions."

"It's a good thing we used only the best materials in the universe. We don't want to disappoint them, do we?"

"Of course not."

"I haven't been to the surface yet," Will mentioned. "How are things progressing up there?"

Mariel suggested, "Why don't we check it out? What do you think, Shanna?"

"Sure. I'd love to."

"You'll get to see some of your friends and their new places of business," Mariel added.

Shanna and Mariel walked slowly with Will and exited the hall. Will limped a little but did so without the use of the cane. He kept one hand against his abdominal incision as a safeguard. Shanna looked worried. "Are you sure you're okay, Will?"

"Yeah, just a little tender."

They ascended a spiral, stone staircase lit by torches. The walls were adorned with symbols and strange letters. One of the steps crumbled under Will's weight. He caught his balance against the wall and complained, "Is this the only way to the top, Mariel?"

"Yes it is, for security reasons."

"How did the Attradeans find the temple and the Eye before?"

"They followed two of our priestesses, who fled from the surface."

"What if they come again? Do you have anything in place to prevent that?"

She smiled slyly. "You'll see."

The top of the staircase was sealed off. Mariel pulled the last torch downward, then released it. The torch eased back to its upright position.

Will and Shanna observed as a tall, narrow section of the left wall retracted inward. They followed Mariel in a single file through the entrance. Once inside a narrow corridor, Mariel pulled on another torch. The section of wall slid back into place.

She walked down the short hallway and stood before a decorative wall. The wall had the carved image of a large eye in the sky looking down at several people. Mariel placed her hand on the image of the eye and dialed it clockwise. A section of the wall rotated a quarter turn, revealing the rear of several shops.

When they exited the corridor to the outside world, Mariel dialed an eye, like the interior one in a counter-clockwise direction and the wall rotated back into place. Will examined the outer wall closely and was impressed. "Wow! You'd never know it was there."

"That's the idea. The other passage was sealed off."

"Does everyone else know about this?" Shanna asked.

"No. Very few individuals are allowed access to the surface and vice versa. Your friends are your responsibility as far as secrecy so be careful who you share this information with."

Will ran a hand across the wall and felt the carvings. "Have any of them used the passage yet?"

"Only Jack, Bastille and Drago, the engineer who designed it. They created this alternate passage. The only access for the others is from the depot on the outskirts of the city."

"So they have to leave the bay and fly to the depot to access the surface?"

"Exactly."

"Good. Let's keep it that way."

Mariel led them onto streets covered in fine, white sand. The shops had cloth exteriors, giving the area an ancient Arabic appearance. The warm, humid air made their clothes cling to their bodies.

Will and Shanna eyed several of the shops. One in particular caught Shanna's attention. There was a variety of ornate clothing items on the racks.

"What do they use for cash, Mariel?" she asked, excited.

"Your friend, Jack, brought in some friends who installed a digital currency system. This system transfers universal credits from non-alliance accounts only so nothing is traceable. The system scans your face and identifies you but doesn't retain the image."

Will looked perplexed. "How do you explain non-alliance accounts to government representatives?"

"There is a second system which links alliance accounts only. The two can't be crossed."

"Isn't that illegal?"

"Not here. This is a neutral planet. We have visitors from all over the universe, both friendly and hostile who need access to either system to make purchases."

"What about Boromians? Can they come here?"

"Yes, so long as they don't bring weapons."

"What about their ships?"

"The ships are tethered at the depot until they are ready to depart. The tether isolates all but auxiliary power on the ship. When the ship departs, it cannot energize its main weapons and tracking systems until it passes a safe distance from the planet. This is done through a program installed in each ship's computer to ensure compliance with Yordic rules."

"Wow! Jack organized all that?" Will responded, astonished.

"Mynx and Neva helped him out quite a bit. It seems they know a lot more than they let on."

"Maya says Jack was always good at stuff like that, too," Shanna said.

"So what about the wedding and the coronation? How do we maintain security?" asked Will.

Mariel explained, "Every visitor who is not a citizen of Yord must register and state their reason for being here. If they are deemed a security risk, they are assigned an escort and given limited access to public areas only. The security force is made up of the women of Yord."

"They are more than capable of handling things. I can attest to that," Shanna added.

Will inquired, "Who has arrived so far that we should be concerned about?"

She pointed across the intersection. "See that shop over there?"

Will and Shanna looked past several passing tourists to a pub. Sitting at an outdoor table were three Boromians. Will was shocked. "What are they doing here?"

"Relax, Will. This is part of our cover."

"I don't get it."

"If they're convinced that there aren't any links to your operations here, they'll spread the word that we're legitimate and leave us alone."

Will became irritable as he eyed the Boromians. Shanna rubbed his arm and said soothingly, "Relax, Honey. Not today."

"What if they find out about our operation?" he questioned.

Mariel warned, "Pray that they don't. We've gone through great pains to camouflage your operation."

Shanna noticed a large shop at the end of the street. She saw blankets and decorative rugs hanging from the eves in front of the store. "Hey, let's go in there!"

"That's Kahlin's store," Mariel mentioned.

"Kahlin's got a store?" Will asked.

"Kahlin is quite a businessman. He buys and sells goods from all over the universe. That's only one of his shops. He owns several."

"I never expected him to trade in his ship for a store."

"He hasn't. It's a front for information and contacts. Your friends, Talia and Zira actually run the store for him."

"I didn't know they were back! Why didn't somebody tell me?"

Shanna tugged at his arm. "Why don't we stop in and see them?"

"I'd like that."

They walked down the street to the store. Talia stood by a table of threads and measured lengths of material for an alien female. Zira pushed her wheel chair toward the counter.

Will called out, "Zira! Talia!"

Talia handed the materials to her customer, who then left the store. Talia's left cheek and nose were bruised. She had stitches in the bridge of her nose and across her chin.

Zira looked up and muttered, "Well look who finally came to say hello." As she approached Will in her wheel chair, he noticed she had a patch of hair shaved from her scalp. The skin was graphed and stitched as were her eyelids, forehead and cheek. Much of it was scarred from the fire on the *Ruined Stone.*

Will felt sympathy for the woman who was no longer the feisty, beautiful girl he once knew. He said apologetically, "I'm really sorry. I didn't know you were back."

Shanna greeted them, "Hi, Talia. Hi, Zira." Zira lowered her head, embarrassed by her appearance and remained strangely silent.

"Why didn't you come see us?" Will asked.

Talia revealed, "We're going through a bad time. When we were discharged from the hospital, we were summoned to an inquiry."

"An inquiry! For what?"

"Zira and I were court-martialed for the *Ruined Stone* incident. We were fortunate that Imperial General Furey only discharged us from the Fleet. The board wanted us executed."

Will couldn't believe his ears. "How could that be? Was Maya there?"

"Yes, she was. She tried to defend us, but the board wouldn't listen to her."

"I guess you know that Maya brought Furey back with her."

Talia's eyes widened with surprise. "What? Why did she do that?"

"Furey bailed on the Fleet. The Weevil have infiltrated the command tier. She's a refugee now."

"Oh, my gosh! So Imperial General Furey may have actually saved us."

"Yes, she did. By the way, she prefers to be called by her first name - Steele."

"You're kidding me. Steele?"

"That's right. So your court-martial was probably the only way for her to get you out of there safely."

"Those bastards crucified us on the stand and they might be nothing but Weevil trash!" she groaned as she broke down in tears. Will tried to comfort her. Talia blurted, "Look at us! We're all busted up! Zira can't even walk."

Will knelt next to Zira's wheelchair and spoke softly. "Talk to me, Zira."

Zira slowly raised her head and looked at him. Tears streamed down her cheeks. "I'm ugly, Will. I'm horrible."

Will hugged her. "No you're not."

Shanna's eyes became moist. She realized what Zira and Talia went through and felt badly for them. Zira sniffled and blurted, "No guy in his right mind would ever look at me now."

Will placed his hand on her cheek and commented, "Zira, it's what's inside that makes you beautiful. Your outer wounds will heal in time."

"That's easy for you to say! You heal fast. I can't."

"Do you believe in me?"

"Why?" she asked him.

"Have I ever let any of my friends down?"

"No. I guess not."

"I don't know what the right thing to do is but we'll figure it out. Have faith in me."

"I'll try."

Will hugged her again and turned his attention to Talia. "So, the two of you are helping Kahlin."

"Yeah, but just until after your coronation. We're part of your security on the surface."

"Do you really think I need security?"

"There are a lot of strangers asking questions about you," Talia replied. "What do we tell them?"

"What do you think you should tell them?"

"I don't know. When strangers asked what the names are of the new king and queen, I tell them I'm new here. That will only fly so long."

"I pretend to be mute," Zira said. "I don't have to answer questions that way."

Will looked to Mariel for guidance. "It's up to you," she told him.

"I never really thought about it. I prefer to stick with my real name, though."

Shanna pondered briefly and then added, "Me, too."

Mariel agreed. "If that's what you like, then it shall be."

"That's not very creative," Zira commented.

"Do you think I need to be more creative, Zira?" Will asked.

"No. You cause enough trouble around here the way you are." They laughed together.

"So our future king and queen will officially be known as Will and Shanna," Mariel declared.

Talia kidded, "It's a pleasure to meet you, your Highnesses."

"Oh, stop it," Will muttered.

"What are you two doing after the coronation?" Shanna asked.

Talia answered, "When I'm fully recovered, I'll join Kahlin on his new freighter."

Zira mumbled sadly, "I don't know what I'll do."

"I'm sure I can use an assistant and a friend," offered Shanna.

Zira perked up. "Really?"

"Sure. We'll have fun. You can stay at the palace with us."

"After all the dumb comments I made, you still want me to be your friend?" Zira asked, surprised.

"Now that you know me better, it's easier for us to be friends." Shanna explained.

"We'll make arrangements so you'll have a room and, of course, a title," Mariel informed Zira.

"That would be great!"

Will's look was somber, displaying his concern for his friends. "Will you be okay in the meantime?"

"We'll be fine," Talia assured him. "Besides, it's only another week in Earth time before your big day."

Shanna kidded, "All I have to do is keep Will from getting hurt."

"You were walking a bit funny when you came in here. What gives?" Talia questioned Will.

"You wouldn't believe it," he said.

"Try me."

"Remember when I was poisoned by the Weevil?"

"Yeah."

"Well, it wasn't poison. Two Weevil pods were growing inside me. I had some problems with dizziness and passing out. Arasthmus ran some tests and discovered the pods. He cut them out and stitched me up."

Talia warned Shanna, "You'd better keep him under wraps. He's already damaged goods."

"I'll try, but it won't be easy."

"We'd best head back before too many people notice us," Mariel said. "You aren't exactly dressed like royalty."

"Can we see the new palace?" Will asked.

"Not yet. It's a surprise and won't be unveiled publicly until the wedding ceremony."

"I don't want to wait. I'm bored," Will complained.

Shanna gave him a coy look. "I can take care of that problem."

Talia and Shanna giggled at Will. He blushed. "Alright, ladies. That's enough." He hugged them and said goodbye.

As Will, Mariel and Shanna crossed the street, three cloaked creatures blocked their path. They slid their hoods back to reveal the horrible faces of Boromians. Their tell-tale dreadlocks hovered above their heads and pointed at Will as if they were poised to strike.

Mariel and Shanna grabbed each other's arms. They trembled but said nothing. Will swallowed hard and asked, "Can I help you, gentlemen?"

One turned on a device around its neck. It spoke in a heavy, frightening tone and asked, "What is your name?"

"Why do you ask?"

"I know you."

"A lot of people know me. I'm Will Saris, the future King of Yord."

Another spoke in a guttural tone, "I've never heard of you."

Will thought quickly and explained, "My future bride and I were raised on Earth in the Solar System. We were brought here on a ship called the *Luna C*. The ship's commander told us we were descendants of the people here and the rightful heirs to the throne."

The first Boromian replied, "I know of the *Luna C*. Where is it now?"

"I believe they returned to Fleet headquarters when they left here."

The three Boromians eyed Will and Shanna suspiciously. One Boromian asked, "What is the name of your queen?"

Shanna courageously stepped forward. "I am Shanna. Who might you be?"

"I am Asheroff, commander of the Boromian forces in this quadrant."

"I didn't know the Boromians had forces in this quadrant," Will remarked.

"Is that a problem for you?"

Will noticed the strain in Asheroff's voice. He replied tactfully, "No, just an observation. What brings someone of your stature to our friendly domain?"

"We are here to assure your intentions are neutral. We've heard that mercenaries hide out on this planet."

Mariel interrupted them, "I am Mariel, the high priestess of Icarus. I have lived here my whole life. If there were mercenaries operating here, I would know it. Everyone here is conducting business. That's what Yord is about – business."

Asheroff stared into her eyes and scanned her thoughts. He recalled that the Attradeans captured priestesses and a sacred, crystal from Yord at one time. Mariel steeled her thoughts to avoid revealing information.

Will interceded, "Asheroff, if you choose to pry into the minds of my people, I'll have you and your party removed from Yord and escorted from the quadrant..." Asheroff snarled, but Will continued, "... however, conduct yourself as a guest and you may stay as long as you like. After the ceremony, I'd be happy to give you a tour of the new palace and its grounds." Will sensed that Asheroff was more tolerable of him after the invitation.

"My apologies, Will. In my position, certain things become habitual. I meant no harm. We do appreciate your courtesy."

"Thank you Asheroff. Will you be attending the ceremony?"

"If my schedule permits me the time. Thank you." The Boromians departed.

"Asheroff is subtly scanning our thoughts," Will mentioned.

Mariel warned, "We need to leave before something goes wrong. That was too close."

They passed the shops and entered the alley in the rear. Mariel turned the symbol of the eye on the wall. The section of wall retracted and they entered the small corridor. She turned the eye on the other side of the wall and it closed.

"They can't read our thoughts now," Will replied confidently.

Mariel sighed with relief. "Good thinking, Will. You handled that well."

"I'm glad you introduced yourself, Shanna. It showed that we aren't intimidated."

"Thanks Will."

Mariel complimented him, "That was a good story you fed them. It'll be difficult for them to validate your lineage on Earth."

"Yeah, but Earth isn't exactly an ally of theirs, especially if Fleet Headquarters is located there."

Mariel pulled on the torch and the next section of wall opened. They passed through and descended the stairwell. Will pondered aloud "Do you think they'll send a wedding gift?" Shanna giggled at the thought.

Mariel was more serious. "I need to coach you on some of the more difficult questions you might run into. We were lucky this time."

"It couldn't hurt, I guess."

Shanna wrapped her arm around his waist. "I think we need to act more like lovers, Will. People should see how much we care about each other."

"Oh, boy," he commented. "I know where this is going."

Mariel laughed. "Perhaps you should slow down a little, Shanna."

"Why does everyone think that?" she groaned.

"Why wouldn't they?" countered Will. Mariel rolled her eyes at the two of them and led them back to the underground facility.

They stopped just outside the main hall when they heard laughing from within. Steele and Jack sat across from each other at a table near the door. They joked repeatedly, while conversing.

Will poked his head through the doorway. "Well, speak of happy couples, look who's here."

Jack snapped, "Keep it up, pal. We already have strained relations, you and I."

Will, Shanna and Mariel entered the hall and sat with them. Will commented, "Nice job on the surface, Jack. I'm impressed."

"You ought to be. While you were playing Sleeping Beauty, I've been taking care of the important stuff – money." Maya slipped into the hall quietly.

"You sound bitter, Jack," Will quipped.

"I am. We're supposed to be friends and we never do anything together anymore."

"What do you want to do?"

"I think we need a road trip."

Maya interrupted, "A road trip where?"

Jack was startled and looked behind him. He was surprised to see her. "Oh, hi, Maya." He put an arm around her waist and kissed her cheek.

Maya asked again sternly, "A road trip where, Jack?"

"Uh, I don't know. Anywhere?" Will and Shanna laughed heartily, enjoying his humbling response to Maya.

Maya asked Shanna, "Didn't I promise you a girls' night out?"

"I believe you did."

Maya suggested to Will, "Maybe you and Jack could help me out and stock up the supply lockers on the *Luna C* this evening."

"I can't lift anything," complained Will. "My side is stitched up."

"I'm sure you two can figure out something." She and Shanna grinned slyly at the men.

"I'll come by for you when we're ready," Maya said to Shanna giddily.

"Do you need me to do anything?" Shanna asked.

"No, I'll take care of the details."

"I'll see you tonight, then."

Jack asked disappointedly, "What about me?"

Maya teased, "I thought you'd never ask. Come with me."

Jack winked at Will. "Sorry, buddy. Duty calls."

Will grinned. "But what about friendship and spending time together? The supplies?"

"That was before Maya's offer."

Will frowned at Shanna. "What's wrong, Will?"

"Nothing."

"I know. You feel neglected."

"No, I'm alright."

Shanna took him by the hand and led him to her room on the upper floor of the base. The room was decorated with floral-patterned wall murals. In the center was a large bed with a velour blanket on top. The blanket had the image of a tiger on it. "I'm impressed," admitted Will. "Does the tiger mean anything on the blanket?"

"Maybe," she replied coyly.

They sat on the bed. Shanna ran her hand through Will's thick, wavy hair and kissed his neck. Will leaned back on the bed and cradled Shanna in his arms. "I guess I know where this is going?"

Shanna chided him, "Oh, shut up, you big baby." As she slid on top of him, they kissed hungrily.

· · · · · · · · ●· · · · · · · · ·

Later that evening, Will and Jack sat on a crate alongside the *Luna C*. They watched as the women approached the dock. Will complained, "I don't like this, Jack."

"Me neither."

Steele, Mynx, Neva, Maya, Talia and Celine met them at the hatch of the *Phantom*. Maya gave Jack a hug and a quick kiss. Meanwhile, Shanna kissed Will hungrily.

Celine teased, "Come on, Shanna. You can do that later."

"I know. I just want to give him something to think about."

"You behave yourself tonight," Will warned her. "We have a reputation to keep up."

Shanna gave him a shy look. "That's why I'm going out."

Will put his head in his hands and grumbled, "This is going to be ugly. I can tell already."

Maya put an arm around Shanna and escorted her onto the ship. "Don't be late!" Jack shouted.

"Be careful," Will added. The women waved and closed the hatch.

Jack sat dejectedly. "Looks like it's me and you partner."

"Any ideas what we should do?" asked Will, bored.

Jack thought for a moment and then responded, "I know a good bar on the surface. Why don't we check it out?"

"Why not? But what about the *Luna C*?"

"Ah, the hell with it. We're not baby-sitters."

Will laughed and said, "That's fine with me but the *Luna C* is your woman's ship. You'll have to answer to Maya when she gets back."

"She'll get over it. Besides, she loves me." They left the dock for the secret passageway.

"You know about the passageway?" Will asked Jack.

"Yeah. Who do you think helped design it?"

"We need to be careful that no one sees us," Will cautioned.

"Yeah, I know. Fortunately, no one goes behind those shops."

"But what if they do?"

"There's a miniature surveillance camera located at each end of the alley. The wall is artificial. We installed canisters of skin irritant in it. So if someone doesn't belong there, they'll subtly get sprayed."

"With what?"

"A paralyzing agent."

"How come we didn't get sprayed?"

"Because our people are monitoring the ends of the alley. They know you, Shanna and Mariel."

"Not bad." Will said. They exited the alley and crossed the street.

"Where is Zira staying?" Will asked.

"In Talia's apartment on top of Kahlin's shop. Why do you ask?"

"I feel bad for her. Let's stop by and see her."

Jack relented, "If you insist."

"She's really messed up and I don't think she's taking things too well."

"Of course she isn't. She's used to being the beauty queen. Now she's Miss Piggy."

"Jack, that's not right."

"I know. I'm sorry," he replied sheepishly.

III

THE FIRST
STRIKE

Will and Jack approached Kahlin's shop and found a group of people standing around the stairwell on the side of the building. "Something's wrong!" Jack exclaimed.

As they hurried toward the shop, Will asked, "What is it, Jack?"

"The shop is closed and the stairwell leads to Zira's apartment. Something must have happened up there."

They pushed through the small crowd. Four women in police uniforms stood at the bottom of the stairwell with energized batons. The tips of the clubs glowed bright red indicating their readiness to discharge, if required. Jack approached the nearest policewoman and called out, "Sylvie, what happened?"

She pulled Jack aside and whispered, "The cripple was killed."

"How?"

"Looks like the Boromians did it. Take a look." Will overheard and became distraught. He tried to push his way past the guards but they held him back.

Jack asked, "Can we see her?"

"Sure. Not much to look at, though." She signaled the other guards to let them pass. Will raced up the steps and entered the apartment.

Zira lay on the floor face up with a vacant look in her eyes. Her wheelchair was broken into pieces and strewn about the apartment. Will shed a tear as he closed her eyes. "This is my fault!" he cried. "I shouldn't have left her alone like this."

Jack tried to console him. "No, it's not. They think the Boromians did it." He swore, "We'll get those bastards. They won't leave this planet alive."

Will examined Zira's face and neck. There were whip marks and red welts in pairs. He touched the wounds and recalled how his mother must have died. He placed his hand on Zira's head and concentrated on her. He could see the attack take place through her eyes.

Three Attradeans attacked her with short crops. One of them opened a small box and took out a small, split-tail scorpion. He set it on her face and taunted it until it stung her repeatedly.

Jack shook Will's shoulder. "Are you alright?"

"Yeah, I'm fine." His voice was filled with sadness.

"I thought you were over this dizzy stuff."

"I am. This is different."

"Like what?"

"The Boromians didn't kill her."

"Then who did? Look at her, Will."

"I did. It wasn't the Boromians. It's a setup."

Sylvie knelt next to Will. She was a big-boned girl with long, dark hair and strong arms. "Did I hear you say this was a set up?"

Will explained, "The Boromians didn't kill her. The stings came in pairs, possibly from a scorpion."

"Now how would you know that?"

He flattened Zira's cheek. "See how straight the welt is. A crop made that mark. If a Boromian dreadlock did this, it would be curved and blistered."

Sylvie inquired, "What are you, an expert on Boromians?"

"Not quite, but I do know some things."

"So how do we pursue this?"

Jack replied, "Suspicious murder. No details available."

"Until…?" she paused.

"Until we say so otherwise. Something's not right about this."

Will stood and slowly walked to the door. Jack thanked Sylvie and followed him.

"The Attradeans want us to go after the Boromians!" uttered Will angrily.

"But why?" asked Jack. "That doesn't make sense." They descended the stairs and passed through the crowd.

"Perhaps it does."

"I guess you want to go back now," complained Jack.

"No, let's go to the bar."

Jack was miffed by Will's decision. "You're serious?"

"Yeah, let's go," Will muttered. They entered the pub and sat at the end of the bar. Will searched the area and saw three Attradeans.

"What'll it be?" the bartender asked.

Jack answered, "A Mind-bender for me."

"I'll have an Iced Tea with hair in it," Will responded. The bartender looked at Jack for an explanation.

"He means that he wants a strong one."

"Oh. Why didn't he say that?" the bartender queried.

"He's not from around here." The bartender laughed and left them.

"What are you staring at, Will?" asked Jack.

"The Attradeans."

Jack noticed the three alien soldiers sitting at a table in the corner. He waited for an explanation but Will continued to stare at them.

"Are you sure you're okay, Will?" Jack asked.

"They killed Zira," he replied, his tone cold.

"So, let's get them."

"No, not yet."

Jack became frustrated and muttered, "Sometimes I just don't understand you, buddy."

The bartender set their drinks on the counter. Jack scanned his face with a palm-sized electronic device and typed in several digits. The bartender thanked him and left.

Will chugged his drink. "We have to go, now."

"Damn it, Will! Make up your mind."

Will stood and glared impatiently at Jack. Jack shook his head angrily, chugged his drink and followed Will outside. "What's going on? I know you're holding out on me."

Will tilted his head back briefly. Jack looked behind them and whispered, "It's the Attradeans. They're following us."

Will nodded and returned to Zira's apartment. The crowd had already dispersed and Zira's body was removed. Jack followed but had no idea what Will was up to.

Will ordered Jack, "Get up on the roof and wait there."

"What are you going to do?"

"Don't worry about it."

Jack grew angry. "Will!" he uttered, his tone gruff.

"I'll tell you later. Trust me."

"Fine, but this had better be good!"

They ascended the stairs to Zira's apartment. Will hoisted Jack onto the roof. He quickly stripped off his clothes and transformed into a wolverine on the doorstep.

Jack leaned over the eave and saw only Will's clothes. "What the hell?" he complained, knowing that Will transformed but unsure why.

The Attradeans climbed the stairwell and examined Will's clothes. The wolverine darted from the darkness and tore out one's throat. It quickly bit into the artery in the thigh of another one and watched as the Attradean fell to the ground, desperately clutching the wound in his leg. His greenish-brown blood poured from the wound, staining the steps. The wolverine lunged toward the last attacker and knocked him over the railing where he fell onto a fence post and impaled himself. The creature transformed back into Will and he quickly dressed.

Jack leapt down from the rooftop and surveyed the Attradeans. He demanded sarcastically, "Would you mind telling me what the hell is going on?"

Will searched the pockets of the Attradean still alive. There he found a tin box, strapped shut. He picked it up and studied it. The Attradean begged, "Please help me. Don't let me die."

"Why should I? You killed a helpless girl."

"I was only following orders."

"Tell me what I want to know and I won't kill you."

The soldier pleaded, "Anything."

"Who is Tenemon going to kidnap?"

"I don't know."

"Come on, Jack. We'll let him bleed to death."

"No, wait!" the man screamed.

Will warned, "Don't waste my time."

"Alright! He's planning to kidnap the Empress Atilena on Urthos."

"That's old news."

Jack was surprised. "You never told me about this. When does this happen?"

"They are making preparations as we speak," answered the Attradean.

Will unhooked the strap around the box and opened it. Inside was the split-tailed scorpion that killed Zira. The Attradean pleaded, "Please! You said you wouldn't kill me."

"I won't kill you but your little friend here will." Will punched the Attradean in the mouth and then dumped the scorpion under his leather vest. He tucked the alien's hand there for good measure. The Attradean's body lurched each time the scorpion stung him.

Jack scowled at Will. "Man, you're in a weird mood tonight."

"We've got to go back and warn Talia. She can't return here." They hurried back to the base and entered the *Luna C.*

"Contact the girls, Jack. We have work to do."

"Alright, but you have a lot of explaining to do."

Will sat down in front of a monitor in the main quarters. He searched the computer database for information on the planet Urthos and Empress Atilena. Jack sat at the communications monitor and signaled the *Phantom.*

Maya's face appeared on the monitor, looking very annoyed. Behind her, Jack was appalled to see scantily clad males dancing. He heard the other women cheering them on. Maya asked angrily, "What is it Jack? This had better be important."

Jack bellowed, "It sure as hell is. Zira's dead and Tenemon is going to kidnap the Empress of Urthos."

Maya shook her head in frustration and looked away briefly. She relaxed and said humbly, "I'm sorry. We're leaving pronto."

"See you soon, Maya," Jack replied but then stomped over to Will and yelled, "They're having a freaking party with strippers!" Will was focused and watched information scroll down on the monitor.

Jack continued in a rage, "Freaking strippers! Do you believe it? And she tells me that this had better be important. What nerve!"

Will placed his hand under his chin and supported his head. He stared at the tabletop. "What's wrong?" asked Jack. "You're not zoning out on me, are you?" Will gestured for him to sit down.

"Please, Will, tell me what's going on," Jack pleaded. Will looked up slowly and smiled. "Well, what is it?" Jack asked, growing more frustrated.

"We've got him again. As soon as the girls get back, we're headed to Urthos. We'll have a slight head start on him."

"Then what?"

"We need an insulated case big enough to put a seven-foot creature inside."

"What in the world are you talking about?"

"The species on Urthos. They don't like humans and they're hostile," explained Will.

"Oh, that's just great!"

"We need four cylinders of compressed Hadyx gas."

"The case isn't a problem, but the Hadyx gas may be."

"We have to have that gas. It's the only way we can kidnap Atilena."

"Wait a minute," snapped Jack. "You said kidnap."

"I did. Now please get the gas. The clock is ticking and we don't have a lot of time."

"Will, you and I have to sit down over drinks and discuss this secrecy bit."

"I promise, Jack. As soon as I have a chance, I'll explain everything."

"And I want you to promise to go on a bachelor party with me and the boys."

"Why?"

"Because the girls did it! So can we."

"Are you getting married?"

"No, but you are. Besides, I'm thinking that I'm better off single."

"We'll discuss it later. Get the gas."

Jack muttered as he headed for the hatch, "Holy crap! Kidnapping. Hadyx gas. What next?"

Will hollered at him, "Easy, Jack. You're getting stressed out."

Arasthmus entered the quarters and asked, "What's all the excitement?"

"Nothing much," Will replied. "We're going to Urthos to kidnap Empress Atilena."

Arasthmus stared at him, wide-eyed for a moment. "Are you out of your mind?" he blurted.

"No, not at all."

"How, pray tell, are you going to pull this off and why?"

"We're going to freeze her, box her up and then transport her onto the *Luna C.*"

"Just like that?"

"Yeah, just like that."

"I'll wait here but, if you pull this one off, I promise I'll wear a woman's dress at the coronation, assuming you survive to attend it."

Will chuckled. "You'd better pick out your dress now. You'll need it."

"We'll see about that. Please, Will, be careful."

"Thanks, Arasthmus." Arasthmus departed the ship.

Will sat down and rubbed his eyes. He thought about Zira and how full of life she was at the dance. He thought about how his feud with Tenemon led to her death. A tear rolled down his cheek as he thought of Shanna and how she could be killed just as Zira was.

IV

CHECKMATE

The *Phantom* glided into its berth and docked. Its hatch opened and the women quickly crossed the deck to the *Luna C*. Inside the ship, Will was asleep at the table with his head resting on his folded arms. The women entered and sat around the table, anxious for an explanation. Shanna rubbed Will's shoulders and woke him.

Will looked around the main cabin and noticed that Steele was missing. "Did you lose somebody?" he asked, curious.

Maya explained that Steele had to take care of something and would join them shortly. "What happened to Zira? Where were you guys?" she asked, concerned. Will told them of Zira's death and the planned abduction of Atilena by Tenemon.

Talia sobbed and shouted angrily, "Tenemon's got to pay for what he did to her."

"I have a plan to kidnap the Empress Atilena before Tenemon does. When his ships arrive, King Rethos' forces will already be on alert. The Attradeans will be sitting ducks when they enter his palace."

Maya asked uneasily, "So what do you need us to do?"

"Nothing. One ship, one abduction, and when the smoke clears, one return."

Maya offered, "I'll go."

"No. The fewer the better, in case things backfire."

"I'll be there with you," Shannon told him, her voice firm.

Will started to speak, but she put a finger to his lips and repeated, "I'll be there with you."

"I'd like that."

Celine asked, "Who else is going?"

"I need one pilot, one navigator and Jack. He'll help me move the equipment."

"I'll go," Celine volunteered.

"Don't you want to check with Bastille?"

"He has to finish his work on the *Leviathan* and I'm bored."

"Take Talia with you, too. It'll do her good to participate in this," suggested Maya.

"Then we're set. As soon as Jack returns, we're leaving."

"Do you have coordinates for Urthos?" Celine asked.

"I already entered them in the computer."

"Wow, you're serious."

"Yes, I am. It's time to take this game to the next level. Tenemon broke the rules and now he's got to pay." Will paced the floor several times and paused. He apologized humbly, "I'm sorry that we had to interrupt your night out."

Shanna asked, "Are you mad at me?"

"For what?"

"We had these, uh…"

Will finished, "Strippers?"

Shanna blushed and said humbly, "Yes."

"No, of course not. Jack on the other hand…"

Maya asked excitedly, "Was he mad?"

Will replied giddily, "Mad is an understatement."

"Good. I owed that jackass."

Jack entered the *Luna C*'s main quarters. "What do you owe me, Maya?"

"Oh, hi, Jack. We'll discuss it later."

He said somberly, "I'm sorry about Zira."

"Yeah, that really sucks."

"We made her assassins pay, for what it's worth," he mentioned.

"Jack, be careful. You guys are playing with fire this time."

"It's no different than Will's last time on Attrades except that I'm in this, too."

"How did you make out, Jack?" asked Will, concerned.

"Funny you should ask. I called in a favor from a friend and got the gas cylinders. The insulated case is outside as well."

Maya was baffled by their conversation. "What kind of gas are you using?"

Jack answered, "Hadyx gas."

"What are you going to do, freeze somebody?"

"As a matter of fact, we are," Will replied, his voice unemotional.

Maya put her hand to her forehead and cringed. "On second thought, I don't want to know any more. Just be careful and come back safe with my ship."

"I'll keep an eye on them, Maya," Shanna assured her. Maya bit her fingernails as she left the ship. Mynx and Neva wished them good luck and followed behind Maya. Shanna sat quietly and brooded over their mission.

"Are you sure about this, Will?" asked Jack, nervous and uncertain.

"Yes, I am. In theory, it will work."

"In theory, we could be killed."

"Yes," responded Will. "In reality, we could be killed."

"Just great. When do we leave?"

"Right now. Let's load the cylinders and the case."

They left the main quarters and stepped outside the hatch. Will surveyed the insulated case, both inside and out. "Nice work, Jack. This is perfect."

"I certainly hope your plan is."

"Me, too." They dragged the white case into the *Luna C*'s main quarters.

"Let's stand it up on the transporter platform," Will instructed.

"On the platform?" questioned Jack, baffled by Will's intentions.

"We're going to use this on Urthos," explained Will. "Relax, Jack. I know what I'm doing."

"That's what worries me. Don't you ever worry that you're due for a fall."

"Well, if I am, we won't be alive to worry about it."

"Oh, well that's just great!"

They lifted the case upright onto the platform then dragged the four gas cylinders next to it. Will closed the hatch and sealed it. He called on the intercom, "Are you ready, Talia?"

"Sure am. Is the hatch sealed?"

"Yes it is. We're all set."

"Okay. Settle in and enjoy the ride."

Jack faced Will, looking antsy, and inquired, "Now what do we do?"

"Get some rest."

Jack uttered irately, "Just like that. Get some rest."

Will chuckled at him and asked, "What's wrong with you? You're a nervous wreck."

"All this stuff you're doing is happening pretty fast. How come I don't know about it?"

"Know about what?"

"About Zira's killers! About the kidnapping conspiracy! About everything! We're supposed to be friends."

Will put his hand on Jack's shoulder and explained, "When I touched Zira, I saw through her eyes how she was murdered. It was horrible."

"I'm sorry, Will. I didn't know you could do that." Shanna got chills from Will's revelation.

Will continued, "Then, inside the bar, I read the minds of the Attradeans. I got a general idea what they were up to. That's why we had to leave."

"Did you know they would follow us?"

"No, but I hoped they would. After you climbed on the roof, I shifted into my alter-shape and attacked the soldiers. We needed one alive to question."

"You never intended to spare him."

"No, he had to die. I'll never get that image of Zira out of my head. I could feel her pain."

"I'm so sorry," said Jack, somewhat embarrassed. "I thought you knew from conventional means and just didn't tell me." Shanna put her arms around Will and leaned against him, saddened by his experience.

"Have you developed any changes since you and Maya have gotten together?" Will asked Jack.

"I never really thought about it. I sense things better, like smell, sight and hearing."

"How about telepathic things?"

"Beats me," replied Jack, embarrassed for not knowing.

"We'll have to work on that. I'm sure you are telepathic and just aren't aware of it."

Jack pondered, "How come Maya never discussed this with me?" When Shanna laughed at him, he grew defensive and asked, "What's so funny?"

"Think about it, Jack," teased Shanna.

Jack became annoyed and responded, "I see, now. She doesn't want me to be her equal."

Will kidded, "Maybe she wants to see how long it takes you to figure out that you are her equal."

"That bitch!" groaned Jack. "Wait until I see her. You guys are going to teach me all about this mind hocus-pocus." They laughed together.

"Only if you promise not to tell her how you found out," Will requested.

Jack shook Will's hand and said, "Deal." He reached for Shanna's hand and shook it, too.

"What's that for?" she asked.

"You won't tell her how I learned this telepathy stuff, right?" Shanna grinned sheepishly.

Will chided her, "You shook on it. That means you have to do it."

"Oh, so that's how this works, huh?"

Will glanced warily at Jack and commented, "Well, usually."

Shanna frowned at him. "What's this usually bit? You do or you don't."

"Okay, you do have to do whatever you shake on," Will replied, knowing he was now obligated to abide by his words.

"Good. Now I understand the rules," Shanna remarked, expecting to use this against the men at some point.

Jack kidded Will, "Your woman drives a hard bargain."

"If you only knew the half of it."

Shanna punched him in the ribs. "Are you complaining?"

He covered his side and warned, "Be careful. The incisions are still healing."

"Then I think it's time for you to get some rest. We can play later." Will and Jack agreed.

Shanna took Will by the hand and led him to one of the cabins. She closed the door and pressed him against the wall with her body. "How is your side, Will?"

"Much better. Why?"

"Oh, I was just wondering."

"About what?"

Shanna kissed him passionately. "Seeing all those male strippers made me miss you very much."

"You don't say."

"Oh, but I do." She pushed him onto the bed and undressed him. Will relented to her, knowing Shanna was not one to be denied.

· · · · · · · ● · · · · · · · · · ·

Will and Shanna sat next to each other in the main quarters at the table. Shanna held his hands in hers. "Will, I want to talk to you about something."

"I'm listening."

"How do you feel about children?"

"I like children."

"No, I mean like children in a family way?"

"Why do you ask?" he inquired, suspicious.

"It's important. We're getting married soon and I want to know how you feel about things."

"I'd love to have a family with you. It would be a little tough explaining to a child why its mother carries knives and dresses in black leather, but I'm sure I'd get through it."

"That means a lot to me." She leaned against him and they kissed.

Will finally asked, "Did something happen to stir your curiosity about kids?"

"Well, yeah. I think I'm... pregnant." She held her breath and waited for Will's response.

Will cracked a smile and remarked, "That would make me very happy."

Shanna was relieved. She wiped a tear of joy from her eye. "I'm not sure, but things are happening to me that seem to indicate I could be."

"That's great, Shanna! We're going to be a family."

Jack entered the main quarters. "What's all this blubbering about?"

Shanna and Will exchanged glances. He pointed to her and replied, "Go ahead."

"We think we're having a baby, Jack!"

"You two are going to be parents? Congratulations!" Jack shook Will's hand and hugged Shanna.

"Thanks, Jack."

Jack hugged Shanna again and responded, "I'm proud of you both."

Celine's voice blared from the intercom, "We're approaching Urthos in cloaked mode. We'll be in position shortly."

Shanna asked Will, "Can I tell her?"

"Sure, but hurry up."

Will activated the monitor and scanned the surface of Urthos. He saw the royal palace and zoomed in. Jack asked, "What do you see?"

"Nothing yet."

Will pressed a switch for the Infra-scan. Three infra-red beams scanned the palace. He zoomed in closer. "See that figure right there?"

Jack looked at a blue shadow on the monitor. "Yeah."

"I believe that's the Empress Atilena. She has no protection detail with her. That's where we're going."

"Now how do you know that's her?"

Will explained confidently, "See this part of the building? Many of the shapes are congregated there. Most likely, that is King Rethos and his sentries. They are most likely in a meeting.

Shanna returned, bubbling with joy, and joined them.

Will continued, "We're landing right here just inside the entrance to the room. That's where we'll spray her with the Hadyx gas."

"Won't that kill her?" Shanna asked.

"No, they use Hadyx gas to anesthetize themselves to heal from injuries." Jack asked, "Will it knock her out?"

"I don't know. The database doesn't tell me everything."

"Oh, that's just great! We're trusting the tabloids of space for information that could determine whether we live or die."

Will kidded, "Details, details, details. Relax, Jack."

"Not out here, I won't. We're not welcome in this part of the universe. If we get caught, we're screwed."

Will patted Jack on the back. In a calm voice, he explained, "I'm well aware of that and I planned accordingly."

Celine's voice blared again from the intercom, "We're in position, Will! The Attradean ships are entering the galaxy at a high rate of speed, too. Watch your time."

"Thanks, Celine."

"Are we ready?" Shanna asked.

Will peered at Shanna. "Still want to come?"

"Of course. You know how I love a good adventure," she replied.

"Jack?"

He answered reluctantly, "Yeah. Why not?" They stepped onto the platform.

Will instructed Shanna, "You open the case immediately. Jack and I will spray the Empress. Just make sure you stay clear of the case front."

"I understand."

"Jack, you go right and spray. I'll go left."

"I got it."

"Okay, Celine. We're ready."

Talia spoke through the intercom, "Good luck, you guys."

Will squeezed Shanna's hand. Jack forced a smile and looked at them as if it was their last goodbye.

On the flight deck, Celine locked in on the target location that Will selected. She activated the transporter and sent them down to the palace.

Will, Shanna and Jack arrived in the corridor outside the Empress' chamber with the tall case and the cylinders of Hadyx gas. It was dark and eerie, giving them shivers. Will muttered, "Uh-oh. Wrong place."

Jack whispered, "Is she in there?"

"I think so. We'll have to flush her out of the room into the hall."

"I have the case door," Shanna whispered. "You and Jack do your thing."

"Are you ready, Jack?" Jack nodded nervously.

"Be careful, Will," Shanna whispered.

"You, too."

Will pushed open the chamber door. He and Jack rushed in with the Hadyx tanks. They were poised, ready to spray the Empress but she wasn't there.

Jack mumbled, "This isn't good." They slowly turned.

The Empress was standing behind them. At almost seven feet tall, she appeared larger when she raised her sickle-like arms. She was a muscular creature with insect limbs and a plain, elongated face. Her skin was more like a shell and was colored dark green and black. Will and Jack trembled as the Empress hissed at them.

Shanna picked up one of the remaining tanks of Hadyx gas and rushed into the room. She sprayed the gas on the Empress from behind and distracted her.

Will and Jack immediately raised the tanks and sprayed the gas on her from the front. Shanna retreated to the corridor and opened the case. "I'm ready, Will!"

Will and Jack moved toward the Empress with the gas spray and forced her toward the case. She had no choice but to blindly retreat inside the case. Shanna slammed the cover and latched it.

Several of the Empress' guards rushed down the hall toward them. Shanna, Will and Jack hid behind the case in a single file, peering out at the approaching guards.

Will yelled into his transmitter, "Now, Celine! Get us out!" In a few brief seconds they were back on the ship. Will attached his cylinder to a port on the case and vented the gas inside.

Shanna giggled. "That was fun."

Will warned, "Don't get cocky. If she gets loose, we're in a lot of trouble." The first cylinder emptied. Will took Shanna's and hooked it up to the port. "Jack, strap down the empty cylinders in the cabinet."

"Roger." He went to the rear of the ship with two empty tanks.

"Alright, Shanna, I need your help."

"What can I do, Will?"

"We're going to try and communicate with the Empress. She may not understand our words so we may have to show her images."

"What do we tell her?"

"Exactly what's going on."

They sat down and focused on the case. At first it was difficult. The Empress was able to resist their attempts to enter her mind. She was also able to filter their thoughts as well in case they wanted to harm her. Will and Shanna portrayed the image of the Attradean ships coming to Urthos and kidnapping her. They showed how they took her instead. Shanna didn't know what to focus on next so she focused on Will.

He showed Atilena how the Emperor Rethus will rally his troops to find his missing spouse. He displayed the Attradeans entering the palace and unknowingly walking into a nest of Rethus' guards, with the Attradeans taken prisoner.

Suddenly, the Empress communicated to Will. "Why have you done this?"

"We are going to return you to your palace. We only took you to protect you from the Attradean forces."

"Why did they want to kidnap me?"

"To force Emperor Rethus to join their alliance and conquer many of the first colony worlds. Would you like to watch the Attradean attempt take place?"

"You would risk freeing me?"

"Yes, I would. How else can I expect you to trust me?"

"Why didn't you just tell us?"

"You don't know us and would surely treat our presence on your world as a threat."

"You are correct. Release me and I will not harm you. You have my word." Will opened his eyes and saw Jack standing nearby.

Will ordered, "Let her out."

Jack was shocked. "You're kidding, right."

Will impatiently got up and approached the case. He unlocked it himself and opened the cover. Jack backed away nervously. The Empress emerged and studied them.

Will inquired, "Do you need time to thaw from the gas?"

"The gas had no effect on me," she revealed. "I'm fine."

Jack and Shanna cast a shocked glance at Will.

He smiled at them humbly. "I'll bring up your palace on the monitor. We can watch what happens." He offered her a seat. The Empress sat awkwardly and watched the monitor patiently. She was cautious and said nothing.

The monitor showed the Attradeans entering the palace from the rooftop. They killed two guards and followed the corridor to the Empress' chamber. Emperor Rethus was there, waiting with ten guards. The Attradeans prepared to battle until more of Rethus' guards trapped them from behind. They quickly arrested the intruders and imprisoned them.

The Empress remarked gratefully, "So you were truthful."

"Yes, ma'am."

"Can I return now?"

"Yes, you can."

"What should I tell the Emperor about his unknown allies?"

"You could tell him that Will Saris, the future King of Yord, wishes him and his Empress well."

"Why don't you return with me? I'm sure he would enjoy hearing it from you."

"I'd be honored, your Highness."

Will looked at Jack and Shanna. "Would you like to join us?" he asked them. When Jack fumbled for words, Will kidded, "Don't sweat it, Jack."

"I'd be honored to meet the Emperor," Shanna answered.

The Empress inquired, "Who are you, young lady?"

Shanna answered proudly, "I am Shanna, the future Queen of Yord."

The Empress studied her for a moment and then responded, "You are welcome as well."

Will contacted Celine on the intercom, "Celine, we're returning to Urthos. All of us but Jack."

"Is the Empress okay?" she inquired, curious.

"The Empress is fine. The Attradeans did exactly what we expected."

"Alright, three of you returning."

Will, Shanna and the Empress stood on the platform. A bright light flashed and then the platform was empty. Shortly after, they appeared in the Empress' chambers.

The Emperor was startled to see them. He hissed and his limbs rose in a threatening motion toward Will and Shanna. The Empress stepped between them and communicated with a series of clicking sounds. She turned to them and ordered, "Follow us."

Guards escorted them behind the Emperor and Empress into an ebony chamber, illuminated by golden radiance from the ceiling. Shanna and Will were amazed by its beauty. Will deduced, "It's probably a mineral in the paint that generates the radiance."

"Well, it's beautiful," Shanna commented.

The Emperor and Empress sat on odd-shaped thrones that were tall and elongated with curved surfaces to accommodate their physical features. Two of the guards positioned a bench in front of Will and Shanna. The Emperor pointed to it. They sat down and waited to be addressed. The Empress introduced them to the Emperor.

The Emperor adjusted a device on his chest and spoke, "You say you are the future King and Queen of Yord. Explain."

Will stood and bowed. "We are to be joined as one in a few cycles. After the marriage ceremony is performed, the coronation will take place."

"I was not aware of such an activity. Our intelligence is usually very good."

"There were individuals who sought to prevent the re-emergence of Yord. This required a degree of secrecy for a while. We would welcome you and the Empress at our ceremony if you are able to attend."

"I was under the impression that Yord had been reduced to rubble and only beggars remained."

"We have reclaimed Yord's glory with her original inhabitants. We seek to make her the economic metropolis and paradise she once was."

"That is a big task," Rethus mentioned.

"We have many hard-working individuals to help us."

"How is it that you knew of the Attradean plot against me?" the Empress asked.

Will related, "Three assassins came to Yord and killed a friend of mine. I was able to read their thoughts and learned of the plot. We captured the assassins and coerced additional details from one of them."

"Why would you go out of your way to help us? You owe us nothing."

"We owe you peace. We owe everyone an opportunity to live in peace and harmony. King Tenemon has fueled his war with new allies by abducting key members of their governments and holding them hostage."

"This was an extremely bold act on your part," the Empress remarked.

"Thank you, your Highness."

The Empress studied Shanna and commented, "You are with child."

Shanna was surprised and responded, "I suspected but wasn't sure."

"Is this your first born?"

"Yes, it is."

"Your child will be wise like her parents."

Will mentioned, "You said 'her'."

"Yes. Your child is a female. She will be healthy and wise."

Shanna replied excitedly, "Thank you, Empress. I am grateful for your insight."

"We must return to our ship now," Will announced. "We still have much to do. I certainly hope you can attend our wedding and coronation on Yord."

The Emperor nodded. "We will come. Thank you for helping us. We are very much in your debt."

"Your friendship is thanks enough." Will then spoke into his transmitter, "Celine, bring us up."

"I assume it's just the two of you," she replied.

"That's correct." Will and Shanna transported back to the *Luna C.*

Jack was relieved to see them again. "It's about time you got back. I was afraid you might have been eaten or worse."

"We had to meet the Emperor," Will explained. "He seems to have taken a liking to us."

"I never heard anything good about the inhabitants of Urthos," complained Jack.

"Whom did you hear that from?" Shanna asked.

"Some merchants told me."

"It couldn't have been too bad if they were alive to talk about it," Will commented.

"Maybe they were exaggerating," admitted Jack.

Shanna called through the intercom, "Take us home, Celine."

Celine asked, "How did it go?"

"Very well. You, uh, missed on your delivery by a little bit."

"I think there's a shield of some sort around the palace. It deflected you a tad off course."

"Well, we made it and they're not mad at us."

"I can't wait to hear about this."

"Oh, you will."

"I can't believe how calm the two of you were when you spoke with the Empress," Jack mentioned to them.

Looking surprised, Will asked, "When, Jack?"

"Here on the ship."

"We never spoke a word," Will informed him.

"Does that mean that you're telepathic skills are developing?" Shanna asked.

"Hey, you're right!" exclaimed Jack. "I didn't even realize it."

Shanna grinned. "So you and Maya really did…"

Jack cut her off. "Stop right there, young lady."

Will chuckled. "What's wrong, Jack?"

"I don't want her to jinx me."

"How can I jinx you?" Shanna asked as she tried to hide a sly smile.

"You never know with Maya." They laughed together.

"Jack, why don't you entertain Celine with the details of our kidnapping?" Will suggested.

"What if she asks about the Emperor?"

"The Emperor and Empress will be at the wedding and the coronation."
Jack's jaw dropped as he stared at them. "You're kidding, right?"

Shanna beamed happily and added, "No, we're serious. They're looking forward to it."

"Oh, boy. Now we're really living dangerously." Jack shook his head uneasily as he ascended the stairs. Shanna took Will by the hand and led him to her quarters.

"Here we go again," Will teased.

"Keep it up, Saris!"

· · · · · · · · ● · · · · · · · · · ·

The *Luna C* moored inside the base at its dock. Bastille and Mariel sat on a crate, waiting anxiously for their friends to exit. When the hatch opened, Will and Shanna emerged first, holding hands. Jack, Talia and Celine followed, smiling cheerfully as they approached Bastille and Mariel.

Bastille, stressed and upset, questioned Will, "What are you doing meddling with Rethus? That's just plain crazy!"

"It's cool, Bastille. Everything went well," Will replied, his voice calm.

"Maya said you were kidnapping the Empress. How could that possibly go well?"

Celine gave Bastille a hug. "Relax, you big oaf. It's fine."

"I guess you're going to tell me that Rethus invited you to the palace to talk about the weather after you kidnapped his spouse."

"Well, yeah," Will answered, kidding.

"Just like that."

"Well, yeah."

Celine, Jack and Shanna chuckled. Bastille looked from one to the other. "What am I missing here?"

Will revealed to him, "They'll be at the wedding and the coronation."

"And the Empress said we're having a baby girl," Shanna added.

Bastille burst into laughter. "You're all crazy. You really did this?"

"They did. I transported them," Celine affirmed.

"Did you see the Empress also?" he asked Celine, feeling left out.

"No, I stayed in the cabin," she replied humbly.

"I saw her," Jack said. "I was shaking in my boots."

"I was frightened at first," Shanna admitted. "But once we let her out of the case, she spoke to us in our language. Then I felt better."

Bastille was shocked. "You let her loose on the ship! Do you know what they eat?"

"Probably a healthy diet of Attradeans right about now," Will replied, chuckling.

Mynx approached them from the *Leviathan*'s hatch. She hollered to Will, "Hey, troublemaker! You have a call."

"Who is it?"

"Tenemon's on the monitor. He's really hot."

"I'll take it on board the *Phantom*."

Will entered the *Phantom* and sat in front of the monitor. Shanna followed and sat next to him. He keyed the receiver and watched Tenemon's angry face fill the monitor. "Good morning, King Tenemon. It's a lovely day in the universe, isn't it?"

Tenemon replied gruffly, "You can hide on Yord but I'll get you. I'm going to kill you, Saris."

"Oh, I see you got the invitation to our coronation,"

"You mean to your funeral!"

"No, I'm talking about the coronation of King Will and Queen Shanna."

"What the hell are you talking about?"

"Why not come and find out?"

"So you'll be hiding behind a throne to avoid me?"

Will laughed at him and answered confidently, "Hell no, Tenemon. You and I go back a long way. I have to say, though, I lost a lot of respect for you when your thugs killed a crippled friend of mine."

Tenemon screamed, "It should have been you!"

"Maybe, but many of the leaders around the galaxy will be appalled when word gets out that the mighty King Tenemon ordered the execution of a poor, handicapped girl."

"Screw you, Saris!"

"Come on, Ten. We've been through this before. It's business and you're not very good at it."

"You're a dead man, Saris!"

"Oh, by the way, Emperor Rethus and the Empress Atilena are looking forward to meeting you at the coronation."

"What?"

"They're enjoying the snacks you sent them." Tenemon screamed as the monitor went blank.

"I think you gave him an aneurysm," Shanna kidded.

"I don't think he's a very happy person right now. He really has to learn to be in touch with his inner self." They laughed giddily and left the *Phantom*.

"I don't think he was aware that we were involved with Rethus and Atilena, do you?"

"No, it sure seemed that way," Will remarked. They left the ship holding hands.

Mariel met them on the dock with a grim look. "We're meeting in the hall. I hope you two can spare a few minutes from your busy schedule."

"For you, Mariel, anything," Will said cheerily. They proceeded up the stairs.

Mariel asked Shanna, "So what happened with Empress Atilena?"

"As far as what?"

"You mentioned you are having a girl."

"Oh, yes. She knew immediately that I was with child. She also said that the baby would be a healthy girl."

"Did she touch you in any way, physically or mentally?"

"We communicated mentally. I felt her scan my thoughts but once she finished, she relaxed and spoke verbally."

"They are a very interesting species, and I believe we have their trust," Will replied. Fearing the rumors she had heard about the Urthonians, Mariel hoped that they were right about them.

When they entered the hall, Will was awed to see his friends and associates at the tables. "Gee, Mariel. I didn't know we had the whole crew here."

"There are urgent issues to be addressed," she informed them. Will and Shanna sat at the head table. Servants entered and brought wine and food to each table.

Mariel stood in front of everyone. She started the meeting by saying, "Thank you all for coming on short notice. We are at a critical juncture with the wedding and coronation in *three days*." She looked at Will for approval.

"Thanks for using time I can relate to," he commented appreciatively.

Mariel nodded and continued, "The palace is nearly finished on the surface. We have several important groups in place to help organize things in our new, growing metropolis." There were claps and cheers.

"From here on out, know that Will and Shanna have been brought here from Earth in the Solar System. This is necessary for the security of our world so we can readily receive protection from all members of our new economic alliance if either of them should be threatened while serving as king and queen."

Will's friends chanted, "Long live Will and Shanna!" Mariel gestured for them to quiet down.

"There are many visitors arriving daily in Yord in anticipation of the wedding ceremony and the coronation. Among them are several suspect individuals. Keep your eyes and ears open for anything strange or threatening."

"Mariel, did something else occur to prompt this meeting other than Zira's murder?" Will asked.

"Many visitors are asking to meet the future King and Queen of Yord. It's time to go public as such." Everyone cheered and held their wine glasses up as a toast to Will and Shanna.

Mariel sat next to Shanna. "After the wedding, it would be appropriate to reveal the news of your child."

Will pulled Shanna close to him and whispered, "I think you should be the one to give everyone the good news."

Shanna was elated and blurted, "I can't wait! I'm so happy."

Mariel expressed her concern. "I'm happy for the two of you, but I'd feel much better if you chose less dangerous tasks in the future."

Will promised, "We'll do our best."

V

THE
CORONATION

The palace loomed three stories high at the center of the city. Inside, the third floor had eight rooms with golden double-doors. Each pair of doors had an ornate lock plated in gold. These were bedrooms for Will, Shanna, Mariel and guests.

The second floor was for the servants. It, too, was decorative with golden double-doors at each room's entrance. The first floor consisted of large meeting rooms, a kitchen, a spacious dining room and a library. There were glass doors leading from the dining room to a veranda at the rear of the palace.

A ten-foot-high wall decorated with mosaic tiles ringed the palace. Two thick, iron gates marked the entrance to the palace grounds. The interior walls were marble and the floors were polished onyx. Beautiful drapes and tapestries covered many of the walls.

Will met Shanna in the corridor of the palace. She wore a long, white gown with a golden train and shawl. Will wore tight, white pants and a black coat with gold buttons and trim. "Shanna, you look stunning," he remarked.

"As do you, my big, strong king." She embraced Will and kissed him.

"This is our big day."

"Can you believe it?" she asked, giddily.

"No, not at all. Just think: by tonight, we'll have been married and ordained, all on the same day. That's quite a feat."

Shanna warned, "I talked with Mynx and Neva. They said to tell you to watch your back. There are people here from GSS, the Fleet and Attrades."

"I see. How about Rethus and Atilena?"

"They're here as well."

"Good. That should slow down the Attradeans somewhat."

Steele and Celine marched down the hall toward them. Celine wore a teal dress with a deep-V middle and a long slit up the side. She greeted them cheerfully, "I had to wish you well before the ceremonies started."

Shanna hugged her and commented, "Bastille's going to have a canary when he sees you."

"Why?" she fretted. "Does it look alright?"

"It's fabulous," Shanna praised her. "You look fantastic." Celine graciously hugged her.

Steele shook hands with Will and hugged Shanna. Steele wore a red, strapless dress with a bare back. Her hair was neatly tucked in a bun. "Congratulations again. I hear you are going to be parents."

Shanna beamed. "We sure are. We're having a daughter."

"How do you know?"

"The Empress from Urthos told me."

Unaware of their recent plot to disrupt the Attradeans, a surprised Steele commented, "Wow! I had no idea you were acquaintances of the Emperor and Empress."

Shanna responded, "It's quite an interesting story how we met."

"The two of you amaze me more and more every day," Steele responded.

Will couldn't help notice how Steele's figure kept her dress perfectly in check. "You look marvelous, Steele."

"Thanks, Will. I know the two of you will be very happy."

"Where have you been hiding?" Will questioned, curious. "I haven't seen much of you lately."

"I've been working with Mynx and Neva on the vaults. I spoke with several friends at Fleet Headquarters and I've got some information for you."

"I hope it's good news."

Steele informed him, "GSS and members of the Fleet have set a trap for you."

"What? Again?" he grumbled.

"Be careful. We'll be nearby to keep an eye on things, but don't let your guard down."

"Thanks," said Will, feeling confident.

"Are you going to announce the new alliance tonight?" she inquired.

"I sure am. Mynx gave me the list of leaders who have pledged to join us. Now, how about some good news?"

"How about this? Ten Fleet cruisers are defecting tonight," Steele revealed. "They'll be positioned nearby in case we have trouble."

"Can you trust them?"

"I've carefully monitored them. They know nothing about the base or our operations, but they are loyal to me."

"Nice work Steele. We'll meet with them when this is behind us."

"I'll be watching your backs. Good luck."

"We appreciate all the help you and the others have given us," Shanna said humbly.

"That's what makes a good team." Steele and Celine then departed.

"I believe it's time for us to be wed," Will declared. He extended his arm. Shanna took it and walked with him down the marble stairs. Maya and Talia waited at the bottom for Shanna.

Will noticed that Jack and Breel were missing. "Where are my best men?"

"They'll be here in a moment," Talia assured him.

Mariel stood impatiently at the entrance to the palace. She motioned for Will and Shanna to come forward. Will searched nervously for Breel and Jack but they were still strangely absent. The blaring of horns signaled the start of the wedding.

Breel and Jack suddenly appeared from the veranda. Jack wore a tuxedo with black shoes. His hair was trimmed and brushed neatly. Breel wore a black leather outfit, contoured to fit his huge frame. The leather had silver trim around the shoulders.

Maya gazed at Jack lustily. She hadn't seen him dressed up and clean-cut since his days at the Academy. It reminded her of why she dated him in the first place. She always found him so attractive and exciting to be around.

"Where were you?" Will asked.

"It's a long story," Breel told him.

Will smelled alcohol on Jack's breath and glared at him. Jack replied innocently, "I'm nervous."

"This is going to be you someday."

"No way. I made up my mind and that's not going to happen." Will laughed at him, knowing he didn't mean it.

The streets were filled with men and women of various races, waiting anxiously to see Yord's first royal marriage in the new era. The bleachers to the left and right of the balcony were filled with honored guests from all over the quadrant. Mariel led Will and Shanna onto the balcony in front of the palace. The crowd cheered until Mariel signaled for silence.

Breel and Jack stood to Will's left. Maya and Talia stood to Shanna's right. Shanna's sister Brindy waited nearby and held a small golden pillow. On top of the pillow was a pair of jewel-encrusted, wedding rings. Four priestesses entered the balcony and stood behind them in a straight line.

Mariel read from an ancient book which held the history of all of their rites. As she read, the priestesses sang with angelic voices. Will squeezed Shanna's hand and smiled at her. She released his hand and pinched his butt.

"Ouch!" he blurted. Mariel paused and glared at him. Shanna held his hand again and giggled.

"Knock it off, brat," he scolded her.

As the ceremony went on, Will scanned the crowd for familiar faces. He noticed one that appeared suspicious, with a face partially hidden by a cloak. As Will studied the face, he recognized King Tenemon. Tenemon caught Will's stare and covered his face. He slid back into the crowd and disappeared.

Brindy presented the rings for Will and Shanna to exchange. They did so and sealed their vows with a kiss. Shanna persisted in making the kiss a long, passionate one. Mariel finally reached for Shanna's arm and pulled her back. Shanna giggled like a little kid. Will blushed as everyone cheered Shanna's enthusiasm. The ceremony concluded and Mariel introduced them as husband and wife. The crowd cheered again.

Will reached behind Shanna and pinched her butt. She jumped and shouted, "Ooh!"

Mariel announced, "The coronation will begin shortly. I ask your patience as our future king and queen make their final preparations."

She turned to them and ordered, "Get inside before I do something very embarrassing to both of you." The newlyweds quickly left the balcony, followed in an orderly sequence by the others.

Mariel chastised Will and Shanna, "That is not the way to behave in a ceremony." Shanna giggled, to Mariel's dismay. Mariel warned, "Young lady, you need to tame those hormones of yours." Will poked Shanna in the side. Mariel then chastised Will, "I thought you knew better than to act as childish as this one."

Will was embarrassed and apologized. "I'm sorry, Mariel. She made me do it."

"Well, stop it! Both of you."

Jack directed his thoughts to Will. "I wish Maya was that playful." Maya crept behind him and grabbed his ear. "Ouch! What's that for?"

"I see you've learned to use your telepathic skills. Just remember, I can read them, too." She pulled Jack into the next room.

Shanna was amused and mentioned. "I love Maya. She is so funny."

"She's not always fun and games," warned Will. "Ask her crew."

"I've heard that's not a problem anymore."

Mariel was annoyed with their childishness. "Return to your rooms and change into your coronation attire before I lose my patience with you."

They raced playfully up the stairs to the third floor. Will entered his room and closed the door. Suddenly, the door swung back and slammed his nose. "Ouch!" he groaned and cupped his hands over his nose.

Shanna stepped in and closed the door. "Come on, Will. Stop messing around."

"What? I'm not messing around."

Shanna looked at his nose and saw it had reddened and swelled. "Oops. Sorry, sweetheart."

"Thanks a lot, Shanna." She took him by the hand and pulled him toward the bed. "What are you doing? We have to get ready."

"There's always time for the important things in our life," she explained with a sly smile on her face. They finished undressing and slid into Will's bed, nestled in each other's arms.

A short while later, Mariel and Jack entered the room. Mariel was shocked to see them under the covers and bare-shouldered. "What in the world are you two doing?" she yelled. "Get up and get ready!"

"It was her idea! I tried to fight her off," Will pleaded.

Mariel rebuked him, "I find that hard to believe."

"There's nothing hard about it, anymore," Shanna teased. Jack burst out laughing. Will pushed Shanna out of the bed with his foot. She barely kept herself covered with the blanket.

Mariel shoved Jack outside and instructed him, "Give them a few minutes. Make sure Will's ready when the music starts." She stormed away angrily.

Will complained to Shanna, "See that. Now Mariel's mad at us."

Shanna quickly dressed and responded, "She's not mad. She's just a little jealous. It's been a while since she's had a man."

"That's mean, Shanna."

"Well, maybe we can fix her up. What about Arasthmus? I think he'd be good for her." Will looked unsure about the idea. "See you in a few, lover." Shanna waved to him and left the room.

Jack entered and asked, "What was that all about Will?"

"Shanna doesn't take no for an answer."

"What happened to your nose?"

Will felt it and winced. "Shanna hit me with the door."

"No kidding."

"Relax, Jack. It was an accident."

"Are you sure?" Jack remarked. He couldn't resist teasing his friend.

Will ignored him and dressed. "So why are you here?"

"Mariel told me to help you get ready."

"Why?" Will asked, looking at him suspiciously.

"She, uh, walked in on Maya and me."

"What were you doing?'

"Nothing yet. We were working on it, though."

Will laughed at him and commented, "So you're just as bad as we are."

"At least I had the courtesy to wait until you were finished."

"Don't blame me, partner. It's Mariel's impeccable timing."

"Why is it that she always catches you and Shanna after the fact and interrupts Maya and me beforehand?"

"It's fate." The two men laughed and tapped their knuckles as a sign of camaraderie.

A short while later, there was a rap on the door. When Jack opened it, Mariel stood there, anxiously waiting. "Is he almost ready?"

Jack turned to Will, "Are you almost ready?"

"Yeah, here I come. I don't feel comfortable wearing a cloak like this," he complained.

Mariel chastised him, "There are many sacrifices you'll have to make for your people. Consider this one of them."

Shanna entered the hallway from two doors down. She wore a long, white gown with lace trim. Over the gown, she wore a heavy, red cloak with white trim. Her hair was tied up neatly and feathered out. Her younger sister, Brindy, escorted her.

Will saw Shanna and was stunned by her appearance. "Wow! You look amazing."

Shanna beamed proudly at him. "Do you like my hair?"

"It's gorgeous."

"Brindy did it for me."

Will praised her, "Very nice, Brindy."

Mariel pressed them, "Come on! Everyone's waiting." She led them back to the balcony.

The priestesses lined up as before. This time, Jack, Breel, Talia and Maya waited in the bleachers. Mariel began the coronation, reading from their book.

Will became nervous as the reality of the coronation set in. Beads of sweat formed on his forehead. Shanna placed an arm around his waist and leaned against him for comfort.

When the priestesses stopped singing, everyone watched and waited quietly. Keira entered the balcony carrying two scepters. Another priestess followed with a crown and a tiara. Mariel read several passages from their Book of Rites. She took the crown and placed it on Will's head. She declared loudly, "I now ordain thee Will, King of Yord." Keira handed him a long, golden scepter.

Mariel then stood before Shanna and placed the golden tiara on her head. She declared, "I now ordain thee Shanna, Queen of Yord." Keira handed her the other scepter.

Will and Shanna approached the front of the balcony and raised their scepters high. Everyone cheered.

Will poked Shanna and said, "You go first."

"Why me?"

"Because you're the comedian."

"You'll pay later, Will Saris." Shanna held her hands up for silence. She spoke nervously at first but then her voice grew stronger. "I'd like to thank all of you for coming today to share in our celebrations. I'd like to make our first official announcement a joyous one: I am with child."

The crowd cheered until she continued, "I have been told by a wise source that it will be a girl." More cheers. "Thank you, everyone." Shanna retreated to Will's side and taunted, "Your turn, hero."

"Thanks, you little brat."

Will spoke loudly and convincingly, "Thank you everyone for participating in the beginning of the new Yord, the economic Mecca of the stars." Applause and cheers rang out.

Will explained, "Yord is the beginning of something new and beautiful in the universe. It is the beginning of an alliance for peace and spiritual development. Through commerce, we hope to draw many races, both human and alien, together in a common, peaceful environment. I've been informed by several leaders that they pledge their support to protect and serve Yord. This is the cornerstone of the new alliance. No longer is war necessary, nor will it be tolerated within our realms."

The cheers grew louder. The crowd chanted, "Long live Yord! Long live Yord!"

Will read from a list of names the leaders pledging support to Yord. Emperor Rethus stood up on the bleachers and attracted Will's attention. Will acknowledged him and asked for quiet.

Emperor Rethus announced in a husky voice with help from an electronic box near his throat, "Urthos would like to join your alliance for peace. We pledge our support."

The crowd cheered and chanted "Urthos! Urthos!"

Will raised a hand for silence and thanked Rethus. He spotted Tenemon in the crowd as he slithered toward the rear of the crowd. Will announced, "King Tenemon of Attrades; you have graced us with your presence today. Would you pledge your support to Yord?" Tenemon froze in his tracks and then turned slowly.

Will pointed to him and asked again, "Would the King of Attrades join our alliance? Don't be bashful." Chuckles were heard from the crowd as everyone waited for Tenemon's response.

Tenemon pulled his cloak back and glared at Will. He responded coldly, "You know I have obligations to an alliance of my own, Saris. I cannot and will not join your alliance."

Will asked coyly, "Pardon my ignorance, but what is this alliance you speak of?"

Tenemon bellowed angrily, "You and I have business to settle. We will settle it soon." He stormed away.

Will kidded with the crowd, "Wow, you'd think I kidnapped his only daughter or something." Everyone laughed at his humor. Breel and Laneia snickered from the front row of the bleachers.

Will finished, "Thank you, everyone. Enjoy the festivities."

More cheers and chants of "Long live Will! Long live Shanna!" floated up from the crowd.

Will escorted Shanna inside the palace where they kissed. Mariel stood nearby with her arms folded, contentedly smiling at them. The priestesses filed past them. Each offered congratulations. When the last priestess left the room, Mariel commented, "I have never seen such audacity."

Will kidded, "What, to marry Shanna?"

Shanna slapped his arm and said, "Hey! That's not nice."

"No," Mariel answered. "I mean to put Tenemon on the spot like that in front of so many people. And then, to kid about taking his daughter away from him as well."

Will replied innocently, "I didn't want him to feel left out."

"He's going to try something, Will," Mariel cautioned. "I don't think you should keep provoking him."

"You're right. I have better things to do with my time."

"Like what?" asked Shanna, curious.

He grabbed her around the waist and tickled her stomach. He lifted her up and spun her around. "Like you, my dear!"

"That tickles. Let me go!"

Mariel interrupted their laughter and ordered, "Stop the child's play! You have to go to the main hall and meet your guests."

Will set Shanna down and grabbed Mariel. He lifted her in the air and spun her around. She laughed uncontrollably as she tried to speak.

"What are you trying to say, Mariel?" Shanna teased.

Finally she uttered, "Put me down, please."

Will set her down but held on to her until she regained her balance. "You've done a lot for us, Mariel, and you've been like a mother. Thank you for everything." He hugged her.

"I enjoy acting like your little girl because you are a mother to me also," Shanna added. "Thank you so much." Shanna hugged her tightly.

Tears welled in Mariel's eyes. She said proudly, "It means a lot to hear the two of you say that."

"Now," Will said, "I think it's time we head to the hall. We don't want to keep our distinguished guests waiting."

Mariel smiled and replied, "Well said."

They left the room and descended the stairs to the first floor. When they entered the meeting room, the members of their new alliance stood and applauded. Will and Shanna proceeded to the head table, followed by Mariel and two priestesses.

Will addressed the gathering: "Thank you my friends. Please be seated and enjoy your dinner." The guests sat down and watched them, anxious to hear more about the new alliance.

Mynx entered the hall and approached Will. She whispered, "Will, I need to speak with you. It's urgent."

Will noted her tone and inquired, "Can we talk here?"

Mariel pulled up a chair for Mynx. She sat down and slid close to Them.

"For some reason, I don't think this is going to be good," he remarked.

"No, it's not. There are eleven Attradean battle cruisers gathering nearby."

"What do you think they're up to?" Shanna asked.

"I don't think they'd be foolish enough to attack, but they surely intend no good."

Shanna fretted, "The *Phantom* and the *Luna C* are no match for eleven Attradean cruisers."

Will pondered for a moment.

"Well, your Highness?" Mynx challenged.

Will grinned, amused by the opportunity this presented. "You know Tenemon is my favorite play toy," he commented. Shanna and Mynx looked confused.

"I think you'd best leave him alone," Shanna warned.

"But he wants a little attention. He's lonely."

"I spoke with a contact of mine by radio and he tells me that the Boromians are amassing in the fourteenth quadrant," Mynx added.

Will frowned. "You know what I think, Mynxie?"

"Yes, I do. The Boromians are going to backdoor Tenemon."

"Exactly."

"What do we do about it?"

"Where's Steele? We could use an appearance by the Fleet's ships."

"I can't find her. The Fleet cruisers have already left the area as well."

"Now why would they do that?" he remarked, curious. "Is the *Leviathan* ready?" he asked.

"Bastille finished the cloaking device today. It's tested and works well."

"After dinner, then, we're taking our guests on a cruise."

"Do you think that's wise?" Shanna asked.

"We don't have a choice. Instruct Maya and Talia to prepare the *Luna C* for flight. Jack will go to man their turret. Celine will take the *Phantom* with Breel on the turret and Laneia as co-pilot. The rest of us will go out in the *Leviathan*. Have Bastille prepare to transport us on board when I give the order."

"What if they attack us here on the surface?" Mariel inquired.

"I'll take care of that," Will assured her. "I want the *Luna C* to take up firing position near the Attradean vessels. I'll reason with Tenemon."

Mynx asked, "What is the *Phantom* going to do?"

"The *Phantom* is to go to Attrades immediately. I want Laneia to find her mother and get her out of there. If something happens to Tenemon, I want her on our side."

"I'll take care of it immediately."

"Thanks, Mynx." She hurriedly from the room.

Will stood and announced to his guests, "After dinner, I have a treat for all of you. We'll be taking a short cruise on a very special ship." The room filled with 'oohs' and 'aahs'.

Servants entered the hall with food and drink for each of the tables. Three of the servants performed a dance with flaming batons in the middle of the hall. They danced to the beat of drums played by two of the other servants.

Mariel was very concerned and questioned Will, "Are you sure you know what you're doing?"

Shanna replied confidently, "If not, I'm sure he'll figure it out."

Mariel fretted, "Things get crazier every day with the two of you."

"And we overcome them each time," Will reminded her. Worried, Mariel sipped from her glass and stared into the crowd.

When dinner was finished, Will walked from the head table and addressed his guests, "Come join us, everyone. There's something I'd like to show you." They exited the palace and stood on the veranda.

Will spoke into a transmitter, strapped on his wrist, "Bastille, we're ready."

A voice crackled back, "How many?"

"Thirty-four."

A flash of light whisked the group on board the *Leviathan*. Mynx awaited their arrival in the ship's main quarters. When they appeared, Will approached her.

"Welcome aboard, Will," Mynx greeted him eagerly. "We're all set."

"Is the *Phantom* under way?"

"Celine took her out immediately. Laneia said to thank you for your concern."

"How about the *Luna C*?"

"They're in position, but Maya thinks they're a little undermanned for the task."

"Hopefully, that won't be an issue. Have Bastille take us behind the Attradean fleet."

Will addressed the guests, "There is an interesting development I'd like for you to witness. I was informed that King Tenemon has eleven cruisers gathering nearby, in hostile fashion."

The guests became uneasy. Emperor Rethus approached Will and offered, "I can summon my warships to assist us."

"Thank you, Emperor, but I have something better in mind. I think you'll like it."

The Emperor responded pleasantly in his gruff voice, "This should be interesting."

Will turned on the monitor over the comm/nav panel. He entered codes for a frequency and waited. An Attradean officer appeared on the monitor. "Who are you? Identify yourself."

Will looked back at his guests with a sly grin. He returned his attention to the monitor and responded, "It's Will Saris, King of Yord. I'd like to speak with my good friend, King Tenemon."

"This is a secure channel. You cannot..."

"Excuse me, soldier, but I do believe your very existence hinges on your delay."

Tenemon pushed the officer out of the way and shouted, "What do you want, Saris?"

"I hate to rain on your parade, but you are stuck between a rock and a hard place, my friend."

"I told you before. I'm going to bury you!"

"Before you get too emotional, I'd better inform you of your options."

"Screw you, Saris!"

"Come on, Tenemon. That's not right."

Tenemon screamed, "I'm going to rain on your parade right about now."

Will grinned, enraging Tenemon even more. "Before you make another blunder, I think you should know that the Boromians are preparing to attack Attrades."

Tenemon's expression froze. "What are you talking about, Saris?"

"I've dispatched a ship to retrieve Laneia's mother as a courtesy to you. The Boromians are expecting your fleet to be crippled by my allies, thus making Attrades easy picking."

"I don't believe you."

Will shook his head in disappointment and taunted, "Tenemon, how did you ever get to be king? You continue to make tactical blunders, one after another." Tenemon's face turned red with rage. Large veins bulged around his head.

"I can give you a demonstration if you like," offered Will.

"The only demonstration is coming at you, Saris!" Tenemon bellowed.

Will informed Mynx to direct Bastille and Maya to target the engines on two of Tenemon's ships and to fire when ready. He turned on a second monitor then returned his focus to the first monitor and spoke calmly, "Well, Tenemon, I see you drive a hard bargain. I have two bits of advice for you. Here they come." Will glanced back at the second monitor. The engines on two of the Attradean cruisers exploded.

On the monitor, Tenemon's face shuddered. He roared at his crew, "Where did those shots come from?" He faced Will and demanded angrily, "What do you want from me, Saris?"

"Just your friendship. Why can't we be friends?"

Tenemon screamed, "Because I hate you!"

"Since you don't understand the gravity of your situation, I'll spell it out for you. So long as the Attradean force is strong enough to counter the Boromian forces, the Boromians won't do anything. If your forces are weakened by my alliance, then the Boromians would enslave or destroy your race. I'm offering to help you."

"The Boromians are part of my alliance. Why would they do that?"

"Look at how you do business. Why would they trust you?"

"Why would I trust you, of all people?"

"It's simple. You return to Attrades and defend your kingdom. We'll provide your wife sanctuary until she can safely return."

Tenemon snarled, "I still don't trust you."

Will shrugged. "I'm trying to help you. What have you got to lose?" The monitor went blank.

Will eagerly watched the other monitor as Attradean cruisers turned back toward Attrades with the two crippled cruisers in tow. Will ordered Mynx to notify Bastille and Maya to return to home base.

"With pleasure," she said.

Will turned to his guests and explained, "This is the *Leviathan*. It is equipped with stealth technology and extraordinary firepower."

Kormut, the Ceratopsian ambassador, was awed. "I heard that all of the Leviathan-class ships were destroyed by a mysterious group of renegades."

Will was amused by the comment. "Yes they were, except for the one that vanished."

Emperor Rethus replied through his voice-mitter, "You've become quite a player, Will."

"This exhibition is an example of how we can manipulate our enemies into a peacetime role."

"That was handled wisely, but what if things had backfired?" Kormut asked.

"If things cannot be handled through peaceful means, we'll respond accordingly. I don't want to put my allies in a combative role, if at all possible. However, when force is required, I pledge to discuss my intended actions with you, my friends, prior to engaging in offensive actions." The guests clapped and cheered.

Shanna added playfully, "The next time we hear from King Tenemon, he may feel differently about things."

"He may not be part of his own alliance anymore, either," Will remarked as everyone laughed heartily.

VI

KIDNAPPED

Will, Shanna and the group of leaders transported back to the veranda outside the palace. "If any of you have questions or concerns, I'd be happy to discuss them," Will announced. "Thank you for joining us and, please, stay as long as you like. You are always welcome here on Yord."

Will and Shanna waved to their guests as they left the veranda and entered the hall. They sat together at a round table near a large fountain. The fountain was centered in a huge rectangular pool with ornate sides. Statues of baby angels were positioned at regular intervals around the fountain. Will hugged Shanna and kissed her cheek.

Shanna remarked proudly, "You handled Tenemon very well."

"Thank you." He removed several of the regal over-garments and handed them to their servants.

"Aren't you worried he might be up to something?"

"It's not Tenemon that I worry about," Will explained. "He's predictable."

"Then who?"

"You, sweetheart!"

Shanna pretended to be shocked and demanded, "How can you say that about me?"

"I'm really worried about what the Boromians are up to. I wonder if the Weevil could be working for them and Tenemon's too naive to know it."

Maya and Talia approached them and sat down. "Well, you made it back here quickly," Will remarked.

Maya complained, "You had me worried. I was afraid you were expecting me to take on that whole armada of cruisers."

"Hell no! The *Leviathan* was ready to take them out if necessary. This was more subtle and Tenemon had no idea where those shots came from."

Maya breathed a sigh of relief and repeated, "I'm not kidding, Will. I was really worried."

Talia hugged Shanna. "Congratulations. You're officially stuck with him."

"Thank you. I'm sure I can handle him," Shanna replied confidently.

"We'll see who handles whom," Will warned her.

"Mind if we borrow your wife for a little while?" Maya asked.

"Of course not. I have to speak with Jack about Tenemon's latest blunder anyway. Have you seen him?"

"He's across the street at the metalsmith's."

"Have fun girls."

Maya added, "By the way, the *Phantom* made it to Attrades using a Fleet portal. Laneia's mother, Queen Siphra, is already aboard."

Will was surprised. "How did they get the coordinates?"

"Bastille figured them out. He believes there's a numeric system that determines the coordinates."

"And he's sure this system works?"

"It has so far."

Will stared blankly as he pondered the possibilities. Maya asked, "What's wrong?"

"If that's true, then anyone could figure out the coordinates for portals."

"Yeah, but I'm sure it's not that easy. Besides, Bastille is a unique individual."

"Some of the alien races are much smarter than we are. What seems complex to us could be much simpler to them."

"Maybe Bastille has a little alien in him," she kidded.

"You never know. Look at us," Will said wryly and walked somberly away.

Maya mentioned to Shanna, "I didn't mean to ruin his night."

"He'll be fine," she replied. "He's obsessed with Tenemon and their feud."

"We're meeting with several of the female guests on the veranda and they requested you to join us," Talia announced. "They'd like to meet you personally."

"Of course. I'd love to."

· · · · · · ·●· ●· · · · · · · · ·

Will left the palace and walked toward the metalsmith's shop. Inside, he noticed several varieties of body armor. Further back was a large selection of swords. He looked about but saw no one. As he glanced at leather uniforms hanging from the ceiling, he suddenly found himself face to face with two Boromian soldiers. The creatures stared coldly at him with vacant black eyes.

Will froze and waited for them to make a move. He tried to read their thoughts but they probed his mind with such intensity that his head ached. He tried to communicate with them but someone struck him on the back of his head and he fell to the ground, unconscious.

A Fleet officer laughed cynically from behind him and mocked the two Boromians. "What was so hard about that?"

The Boromians snarled angrily at the officer. They dragged Will out the back door of the shop into the alley where Jack and the metalsmith lay unconscious on the ground.

The Fleet officer ordered the Boromians, "You stay here and keep an eye on his friends." He took a remote device from his pocket and pressed two buttons. Will and Jack were transported onto a Boromian vessel along with the officer. Two more Boromian soldiers promptly dragged them into a sophisticated laboratory. They laid them on tables and strapped their arms and legs to the sides.

Will's head spun as he regained consciousness. He opened one eye slightly and saw Jack lying on the table next to him. He used telepathy to communicate with him. "Jack, can you hear me?"

"Not so loud. My head hurts!" he responded.

"Don't think about anything they could use against us. They'll read our thoughts."

"When does that happen?"

"As soon as they realize we're conscious."

"How do we get out of this?"

Will instructed him, "Keep still."

Two Weevil entered the room, followed by two Fleet officers. They surrounded Will and studied him. The first officer remarked, "So this is the elusive Will Saris."

One of the Boromians mocked, "Now he's the not so elusive King of Yord."

The officers stared angrily at the Boromians. The first officer inquired, "He's been ordained?"

"Yes, he was. But that didn't help him."

"You kidnapped a king? You buffoon!"

The Boromian replied, "One of your officers led the abduction."

The second officer said, "I know who did it. I'll deal with him." He left the room hastily.

The Boromian asked, "What's the big deal?"

The officer explained, "Saris has a lot of friends out there. If he was ordained king, his allies will come looking for him. Now is not the time for distractions to our operation."

The Weevil asked the Boromian, "Have you located the Fleet's commander yet?"

"No. She vanished and hasn't been seen or heard from."

"Hmm. That could be a problem."

The Boromian assured him, "We can handle this group of renegades."

The Weevil explained, "If Furey gains control of the Fleet's ships before we do, then we've got a problem. "You still haven't located the missing destroyer either."

"We think the key to locating it is in Tenemon's armada. When his forces attack, the renegades will have to bring *Leviathan* out of hiding or risk having their empire decimated."

The Weevil asked his partner, "Have we heard anything on the attack yet?"

"No, but it should be soon."

· · · · · · · · ● · · · · · · · · · ·

Tenemon's armada raced back to Attrades. His palace was in flames and Boromian battle cruisers patrolled the skies.

The Attradean cruisers surprised the Boromian attackers and engaged in an intense battle. After several hours of fighting, only two Boromian battle cruisers limped away, both crippled.

Tenemon's cruiser was the only remaining Attradean ship left intact. He flew over the major cities of Attrades, surveying the death and destruction.

"Perhaps we should declare a truce with Saris," his lieutenant suggested.

Tenemon barked, "For what reason?"

"We need protection. Our people need help."

Tenemon hung his head in shame and groaned, "I've failed my people."

"What do we do next, my king?"

"We need to help our people first. Set the ship down."

"Right away, sir."

"Send a message to Yord. Tell them what has happened and ask for their mercy as well as any assistance they can send."

"I will, sir."

"Oh, and ask if they have Siphra in their custody."

"Yes, sir." The officer departed the main quarters.

The Attradean ship landed outside the palace grounds. The crew exited and searched the area for survivors. Tenemon stood in front of his ship and stared at the burning ruins of his palace, wondering how it came to this. Once, he was the most feared leader in that part of the universe. Now, his force was reduced to just one ship and his kingdom was decimated. He had become the thing he feared most - insignificant.

· · · · · · · · ● · · · · · · · · · ·

The *Phantom* sped through the second portal. Celine received a beep for an incoming signal and activated the monitor. The face of an Attradean officer appeared. Celine taunted him, "Well, well, well. To what do we owe this?"

"I'm calling on behalf of King Tenemon. Boromian ships have ravaged our cities. We destroyed all but two of them. We, ourselves, are the only cruiser left and we desperately need help."

"So what do you want from us?"

"Look, we all know that Tenemon isn't a smart or trustworthy leader but I think he has paid the price. We all have."

Celine put him on hold. She spoke into the intercom, "Laneia. Breel. Come to the flight deck immediately. I suggest you bring Queen Siphra as well." She turned her attention back to the monitor and pressed 'resume'. "So what do you need from us?"

The officer explained, "We have many wounded. The cities and palace are burning piles of debris. Can you get help for our people?"

Laneia, Breel and Siphra entered the flight deck. Siphra recognized the officer and asked, "How bad was the attack, Mantos?"

The officer's eyes widened with surprise and he exclaimed, "Queen Siphra! You're alive!"

"Of course I am, thanks to my daughter and her friends."

"It's bad. We have but one ship left to fight with. All our cities are gone. Casualties are catastrophic."

Siphra looked at Celine. "Can you do anything for them?"

"We'll discuss it with Will and Shanna. It's going to be their decision. Tenemon hasn't exactly made friends around the universe these days."

"I'm aware of that. I can take responsibility for what will happen to him."

"How would you do that?" Celine asked curiously.

"He's broken his vow to protect and serve his people. As a result, he must step down from his throne. I can retain my throne and become the dominant leader of Attrades as a result of his failure."

"I like that. Now the female gets to wear the pants."

Siphra advised Lieutenant Mantos, "I'll meet with the leaders of Yord. I can't promise anything, but I'll see what I can do."

"What should I tell the king?" Mantos asked.

"Remind him of the consequences of his actions. I'll be in touch."

"Thank you, your Highness." The monitor went blank.

Laneia hugged her mother. Siphra placed her arms around Breel and Laneia. Her voice was tinged with sadness, "Maybe this is what Attrades needs - a new beginning. It's a shame it had to come at such a high price."

Celine contacted Mynx on the monitor. She answered from the *Leviathan*. "How did you fare on Attrades?"

"Very well. We rescued the Queen just before the assault. The Boromians attacked, as we expected, and Attrades was demolished."

"What kind of casualties did they inflict on the Boromians?" Mynx inquired, concerned.

"Sounds like two Boromian ships escaped and only Tenemon's cruiser survived the battle."

"Wow! So the alliance is up for grabs. I wonder who's going to step up to the plate now that the two main contenders are beaten up."

"Good question. Where are Will and Shanna?"

"At the palace. What's up?"

"Tenemon's no longer in charge. Queen Siphra is. They need our help on Attrades."

"I'll notify them immediately and get back to you." Mynx changed channels and tried to raise Will via his transmitter but got no response. She tried Shanna and waited anxiously.

Shanna answered. "What's up, Mynxie?"

"I just talked to Celine. Attrades has been ravaged. Queen Siphra is now in charge and they need our help."

"I'll find Will and call you back."

"I tried to contact him but got no answer."

"Maybe he's with Jack. I'll get right back to you."

Shanna excused herself from the table, to her guests' disappointment. "I'm sorry, ladies. Something urgent has come up. I'll be back in a while."

She stepped outside the palace onto the veranda and called Will several times on her transmitter but still received no response. She changed the code and called Jack several times, with the same result. Shanna fretted and called Maya. Maya answered, "Go ahead, Shanna."

"We've got problems. I need you, Talia, Steele, Mynx and Neva right away. I'll meet you on the third floor."

"Did you talk to Will?"

"I can't contact him or Jack. I'm afraid something has happened to them."

"We'll be there shortly."

Shanna hurried to the stairwell and met Arasthmus. "Well, good evening, your Highness. Do you have a few moments?"

Shanna replied emotionally, "Not now, Arasthmus. We have big problems."

"What kind of problems?"

"Attrades was demolished by the Boromians. Tenemon's forces were destroyed and they are asking for assistance."

"What's so bad about that?"

"Will and Jack are missing!"

"Don't panic. Maybe they're occupied."

Shanna shouted, "I can't sense them either! They aren't on Yord anymore."

When she became teary-eyed, Arasthmus comforted her. He reminded her, "Remember, you are a queen now. You have to keep your composure."

"I'm sorry. It's just that I'm really worried about them. I can sense they're in trouble."

"What have you done so far?"

"Maya and the other women are on the way. I hope they can help me decide what to do."

"Sit down and relax. Think about what is going on and evaluate your options logically. Decide what your immediate priority is." Shanna took a deep breath and tried to relax. Her heart raced as she tried to narrow her thoughts down to one.

He asked, "What do you think is most important right now?"

"The people of Attrades need help. But what can we do?"

"Concentrate on your resources."

Shanna thought about everything that happened at the ceremonies and who attended. "I've got it! I can talk to our guests. They have the ability and the resources to help Attrades."

"Now you're thinking like a queen."

"But what about Will and Jack?"

"For that, you'll need help from your friends."

"Thank you, Arasthmus."

"My pleasure. Before everyone gets here, I need to tell you something."

"Yes, Arasthmus."

"I think I found a way to beat the Weevil."

"That's great, but it won't help us now."

"I understand. We'll discuss it later."

"I'm sorry, Arasthmus, but I'm really worried about the boys."

"I would be happy to assist in whatever way I can."

"Good. Stay with me while we figure out what to do."

Mariel entered the hall from the stairwell. "What's going on, Shanna?"

"Will and Jack are missing!"

"What! I'll see what I can find out from the Eye." Mariel hurried out of the room.

The other women entered the hallway. Shanna motioned for them to follow her into the master bedroom.

"Any word from Will or Jack?" Maya asked Shanna.

"None. I wanted to consult with all of you because you are my friends and I trust your judgment." They were eager to offer their support. Arasthmus stood back from the group and listened patiently.

Shanna explained, "I'm going to ask our guests for help in aiding the people of Attrades. The Boromians have left their cities in ruins."

"Why should anyone care about Attrades?" Maya asked.

"Because Queen Siphra is in charge now. Tenemon is out of the picture and I seem to remember from talking with Will that Attrades was the center of the alliance. That's a critical power void for the Boromians and the Weevil to fight over."

"Makes sense to me," Talia commented.

"Diplomatically, it could give us an edge against our enemies," added Mynx.

"I think it's a wise move," Steele said.

"Personally, I think we should take the *Leviathan* out in preparation for another attack of some kind," Shanna told them.

"It's possible the next attack could be on Yord," Maya suggested.

"We only have three ships. We'll have to use them discretely."

"I've been in contact with close friends on Fleet ships," Steele informed them. "I now have the support of seventeen Fleet battle cruisers at my disposal."

"We wondered where you disappeared to," Mynx commented.

"Timing was critical to secure their allegiance."

"How do you suggest we use them?" Shanna asked.

Arasthmus pointed out, "This is similar to a game of chess."

Steele chided cynically, "This is no game, sir."

"Ah, but it is. Let's think about this. What happens if everyone diverts his or her attention to Attrades? What if they are really trying to bait us into revealing the *Leviathan*?"

Looking disappointed, Steele remarked, "We already have questions. How about answers, old man?"

Arasthmus ignored her slur. "Let Rethus and Atilena handle Attrades," he suggested. "I'm sure they'd be happy to negotiate a truce through Siphra rather than Tenemon. Now you have the other allies' forces at your disposal." Maya agreed. Mynx and Neva consented as well.

He addressed Steele. "You say you have seventeen ships at your beckon?"

"Yes, I do."

"How do you feel about saving the Fleet, Steele?"

"If I could, I would."

"Then I think you should take those ships and recover your headquarters."

"And what would I gain by doing that? I still don't know who is Weevil and who is human."

"What if you could?"

"And how would I do that?" she asked, still sarcastic.

Arasthmus reached under the long flowing sleeve of his robe and retrieved a spray bottle. He handed it to Steele. Everyone eyed the bottle, curious as to its contents.

"What is it?" she asked.

"It's something I developed after studying Shanna's blood sample and the DNA from the pods. If you spray it into the face of a Weevil imposter, he will convulse, thus making him defenseless. It does nothing to other species. After that, it's up to you what you do with them."

Steele's attitude warmed. "How much of this have you got?"

"A half dozen bottles."

"And you're sure it works?"

"I'm quite confident it will do the job. I've identified the protein in Shanna's blood that repels or destroys Weevil DNA. Once ingested, it should mess them up pretty good."

Steele held the bottle up smugly. "Nice work, Arasthmus. Now it's my turn to kick some Weevil ass."

"I think we need to have the Fleet's resources readily available," he suggested. "Right now, we don't know who will take over the alien alliance and what their motives will be."

"Maya, I think you and Steele should lead the attack on Fleet headquarters," Shanna suggested. "Once the Fleet is safely restored under

Steele's control, we can turn the tide against our enemies in an organized manner."

Mariel rushed into the room and announced, "We know that Will and Jack are on a Boromian ship! The Eye showed them as captives."

"I should have focused on the Eye sooner," Shanna fretted.

"You'll learn, young lady," Mariel chided. "The Eye is a valuable resource."

"Where are they?" Shanna asked.

"We don't know for sure. If we keep focusing on the Eye, it may give us a lead."

Shanna sighed and looked down at the floor sadly. "Are they okay?" she asked, fearing the worst.

"So far as we can tell."

Steele praised her. "You've done well, Shanna. If Arasthmus' spray works on the Weevil, we'll be back soon to help you find Will."

"Thanks, Steele."

They left the room and proceeded to the transport bay. Shanna and Celine hurried to the *Phantom* and summoned a crew.

· · · · · · · · ● · · · · · · · · · ·

The senior Fleet officer shook Will several times. Will finally opened his eyes and acknowledged him. "Fancy seeing you, an officer of the Fleet, here with Boromians and the Weevil."

The officer slapped him and admonished, "You simpleton! I am a Weevil."

Will replied sarcastically, "No kidding, Sherlock. Did you really think you had us fooled?"

"It doesn't matter. Soon we'll rule another quadrant and one day, the universe."

Will looked at the Boromians and commented in mocking fashion, "Initially, I thought you guys would have been running the show, not these losers." The Boromians snickered.

The officer angrily punched Will but he continued to taunt him further, "If you think that makes you a man, forget it."

The officer grabbed Will's chin and squeezed it tightly. "Enough games, Saris. Where is Furey?" he demanded.

Will was surprised by the question. "Furey? What about me?"

Jack goaded the officer, "What's the matter? You need a date or something?"

The officer turned to Jack and punched him in the stomach. A glob of saliva flew from Jack's mouth and struck the officer in the face. He licked it off and held his hand up. Sharp, pointed claws emerged from the fingers and then drops of clear liquid beaded on the tips and rolled down each finger.

"I think you'll make a very good Weevil."

Will laughed at the creature hysterically. Jack wasn't sure why but he followed Will's lead and laughed too. The officer shouted, "What's so funny?"

"You really aren't that smart, are you?" Will asked. The Boromians snickered again.

The officer responded angrily, "Then you shall be the first!" He turned toward Will with claws extended. Will laughed harder at him until the officer held his claws to Will's throat. Jack continued to laugh, although he still wasn't sure why.

Another officer cautioned his superior, "Wait. Something's wrong here." He approached Will and leaned over him. "Maybe we aren't as smart as we think. What's so funny?"

Will answered, "If he sticks either of us, he'll be in for a big surprise."

"Like what?"

"Like he'll die a horrible death. We have been vaccinated with a serum. If he tries to inject us, he'll explode."

The officer laughed at him and responded, "That's pretty lame."

"Oh, you think so? Look at the scars on my shoulder. You think you're the first one to try me?"

The officer pulled Will's shirt collar back, exposing his shoulder. There were five scars from puncture wounds.

"If you check the back side of my shoulder, there's another set of scars," Will revealed.

The senior officer asked, "How could you live through two injections?"

"Easy. The vaccinations work quite well."

"I don't believe you!"

Will provoked him further, "Then go ahead and try, you fool. I dare you."

"All we want from you is the location of Steele Furey. You know where she is."

"Why would we hang with Furey?" Jack asked. "She kicked me out of the Fleet a long time ago."

"And I quit the Fleet because she was clueless. That's why we were losing the war," Will added.

"Then I guess you're of no use to us. We'll leave you to the Boromians," the officer said, wryly. He motioned for his subordinate to follow and approached the door.

Will warned, "Watch your back, boys. We're coming for you."

A Boromian entered the room. He ranted wildly, "The attack was a disaster! We lost all but two of our ships."

The senior Weevil asked calmly, "How is it that you botched this?"

"The Attradeans came back! They knew we were coming."

"So why did your last two ships run? They could still fight."

"They were badly damaged. They had no firing capability."

"Why didn't the Attradeans pursue them?"

"They had but one ship left themselves."

"Good. Then it's time for us to take control of the alliance."

"But the alliance is supposed to be ours."

"And you'll defend it with what - your worthless fighting force?" the officer challenged. The other Boromians snarled at him.

The officer drew his pulse pistol and shot each of them. He sneered at Will and Jack, "I guess you can stay here and die, unless you happen to remember where Steele is."

Jack commented cynically, "The Weevil aren't very intelligent, are they?"

"No. They're obviously much more primitive than we thought," replied Will.

"It's just a matter of time before they're eradicated, like a disease."

The officer punched Jack in the face and warned, "We'll see who gets eradicated." He and his partner left the room.

With some difficulty, due to his jaw aching, Jack asked Will, "Do we really have a vaccination to stop the Weevil?"

"No. I just made it up."

"Well, you had me convinced. What if they would have stuck us?"

"Then Arasthmus would be digging pods out of us again."

"Well, that's just great."

One of the Boromians crawled across the floor, barely alive. Will noticed and used his telepathy to communicate with the creature. Jack listened, wondering if the creature would help them.

The Boromian pulled himself up against the table. It snarled at Will but unbuckled the straps on his wrists. The creature hissed and then collapsed on the floor. Will quickly unbuckled the straps on his ankles and unbuckled Jack's wrists.

"Why are they hell bent on finding Steele?" Jack asked.

"I'm not sure. Maybe she knows something about them."

Jack unbuckled the straps around his ankles. "Do you think she's one of them?"

"No, but maybe she has a vested interest in what's going on."

They tried to leave the room through the hatch but it was locked. "This is ridiculous. Now what do we do?" Jack asked.

Will looked at the Boromian corpses. They still had their weapons on them.

"The Weevil must be pretty stupid to leave the Boromians' weapons on their corpses," Will mentioned.

Will and Jack took two pulse pistols each. They pointed them at the hatch and fired repeatedly. The door heated up until it glowed red. They continued firing until the door turned white hot. "That's enough," Will said.

"Now what?"

Will looked at one of the dead Boromians. "Grab the feet."

"For what?"

"For getting out of here!"

Jack reluctantly grabbed the feet of one of the corpses. Will lifted the corpse by its wrists. "Now we throw it against the door," he explained.

They counted together as they swung the corpse back and forth. "One. Two. Three!" They tossed the body against the white-hot hatch. The hatch broke away from its hinges and fell outward. The corpse sizzled from the heat of the door.

"How did you know that would work?" Jack asked.

"Did you have a better idea?" countered Will.

Jack looked around the room, "We could have used the table."

"Yeah, we could have, but we didn't." They carefully stepped past the smoldering corpse and molten door.

"Where to next?"

"To find a radio and call for help," Will replied.

At the end of the corridor, Jack saw a stairwell. "This way," he called out. Will followed him.

At the top of the stairs, they found the entrance to the Boromian control room. They opened the door slightly and peered in. Seven Boromian officers lay dead on the floor. They cautiously entered the control room.

"This definitely isn't good," Jack uttered.

"I'll search for survivors. You find the radio and get an SOS out."

"Be careful, Will."

"And you watch your back." Will left and descended another set of stairs.

Jack found the radio and tried to send out a distress signal. He made several attempts but got no response. His frustration mounted as he repeatedly tried to contact Maya, Celine or Bastille.

Will heard alien voices from the lower level as he descended a second flight of stairs. He crept along the corridor, passing several doors on the right. The fourth door was the entrance to a brig. The door had iron bars on its window.

Will peered inside and saw what was left of the Boromian crew bound on the floor. One of them recognized him and called out, "Saris. Will Saris."

Will looked again and inquired, "Who is that?"

"It's Asheroff."

Will was surprised. "How is it that the Boromian commander is locked up in the brig of his own ship?"

"We've all been duped by the Weevil. They are preparing for a major invasion."

"Who would they invade? You? Attrades? Yord?"

"All of us. By weakening our forces against the Attradeans, we made ourselves vulnerable."

"Do you think we could coexist instead of killing each other?" Will asked him.

"By nature, Boromians are warriors. We realize we could all be facing extermination if we don't work together."

"What do you propose, Asheroff?"

"I think your group could use us to defend your alliance."

"Like mercenaries?"

"Exactly. We don't have families like other races, nor do we have homes. We're just nomadic warriors."

Will pulled on the door's handle. It was unlocked. He opened the door and approached Asheroff, wary of the Boromian's poisonous dreadlocks, but they never stirred. He untied Asheroff and helped him to his feet. "Do you know where the Weevil went?"

"They are part of an advanced task force. Their mission is to undermine alliances for a bigger invasion later." He helped Will untie the other Boromians.

"You could come back to Yord with us. We can discuss the situation with some of our other allies," Will suggested.

"I'm surprised you would trust us, Saris."

"We have to start somewhere." They shook hands.

Will explained, "My friend, Jack, is trying to contact our people. I need to see if he's had any luck."

They ascended the stairs to the top floor. Seven other Boromian officers followed. Jack was shocked to see Asheroff with Will. Will asked, "Did you make contact yet?"

"Not yet."

Asheroff informed them, "You won't be able to contact any of your ships without disabling the filter." He approached the control panel and activated the monitor. He pressed the monitor control switch four times, scrolling through menus as he did. He tapped a button to the left of the monitor and the monitor blinked. "Try it now." Jack entered the frequency again and sent another message.

On board the *Phantom*, Celine and Shanna manned the flight deck. Shanna asked despondently, "Where do we even start to look for them?"

"We can head toward Attrades, since that's where the Boromians were last seen," Celine suggested, watching the monitor closely. "The long-range sensors will tell us if there's a Boromian ship in the general vicinity. But, if they're on the other side of a portal, we'll never know."

The monitor beeped three times. Shanna and Celine stood up anxiously. Celine glanced at the display and mentioned to them, "I don't recognize the source of the transmission." She keyed the receiver and responded, "This is the *Phantom*. Go ahead."

When Jack appeared on the monitor. The girls' faces lit up with hope. Celine yelled, "Where are you, you jackass?"

"Easy, girl. We're on a Boromian flagship."

"Where's Will?" Shanna asked, her concern evident in her tone.

"He's right here."

Will's face appeared on the monitor. "Listen up, Celine. The Weevil ambushed the Boromians. Without the Boromian and Attradean forces, the quadrant is wide open for the taking and the Weevil look like the aggressors."

"Where are you?" she asked.

"I don't really know. Where's Maya?" he queried.

"Keep talking," instructed Celine. "I'm trying to trace your signal."

"Arasthmus developed a spray that exposes the Weevil," Shanna informed them. "Maya and Steele are leading a group of Fleet ships to recover their headquarters."

"Excellent. Contact them and warn them that the Weevil are behind everything," Will ordered.

"I've got a lock on you," Celine announced. "We'll be there soon."

"I can always count on you girls," Will replied, thankful for their efforts.

"Ditto." The monitor went blank.

"Those two clowns always seem to dodge trouble," Celine complained.

"I can't imagine how they made peace with the Boromians in the first place," Shanna remarked. "This should be an interesting story."

"Aren't they all?" Celine quipped.

· · · · · · ●●●●● ● ●●●●● · · · · ·

Two Boromians sat at the controls and initiated several systems checks. One of them called out urgently, "Asheroff! Look at this."

Asheroff approached and asked, "What is it?"

The Boromian pointed to the monitor. Asheroff saw three Weevil craft in a firing position. "Force fields up, quickly!" he ordered. The ship rocked violently.

"Who's manning the cannons?" Asheroff bellowed.

"Scinter's got it."

"Well, he'd better do something quick." The ship rocked again. A loud siren wailed.

"What's that?" Will asked.

"Loss of auxiliary power. Our shields are gone," Asheroff replied.

Scinter's voice echoed from the page, "I've got no power! They're closing in."

Asheroff groaned despondently, "We're as good as dead."

They stared at the monitor. A single, green dot approached rapidly from the top right corner. "What's that?" Will asked.

Asheroff exclaimed, "Someone's coming at us fast!"

The ship rocked violently to the left as the men held on desperately to the control panels. Six smaller, red dots broke away from the larger dot and fanned out toward the Weevil ships.

Asheroff yelled excitedly, "Six torpedoes heading toward the Weevil ships! Veer right at six kleegs," he ordered his crewmen. Get us away from here." Another blast rocked the ship.

They watched the dots draw closer to the Weevil. Four of the dots disappeared when they intersected two of the three ships. The ships became bright red spots and then disappeared. The remaining ship turned to confront the attacker. A second wave of six torpedoes already approached the Weevil craft.

"I don't think they see the torpedoes coming," Will remarked, hopeful. "They're heading right into them."

"Maybe Bastille found a way to cloak them, too," Jack suggested.

The Weevil ship fired two torpedoes just before impact. A few seconds later, the ship disappeared from the monitor, just like the first two. They watched the approaching ship dodge the torpedoes as it drew near.

Celine's voice crackled from the speaker, "This is the *Phantom*. Please respond."

"That's Celine!" Will shouted.

Jack hurried to the radio and answered excitedly, "Celine! It's us."

"Are you guys okay?"

"We are now."

"Can we follow them?" Will asked Asheroff.

Asheroff looked at his crewmen. They replied in their native language. Asheroff explained, "We can follow but only at half-speed."

"Tell them to stay close," Will advised. "We're all going back to Yord." Jack relayed the information to Celine.

"We owe you our lives," Asheroff said humbly.

"Perhaps we can repair your ship when we get to Yord."

"Do you have a base there?"

Will answered carefully, "We have skilled technicians there who can help you."

"That will suffice. We owe the Weevil for their treachery."

Will replied assuredly, "We'll all get our chance at the Weevil. I promise."

The Boromian ship limped toward Yord with the *Phantom* nearby for cover.

VII
BATTLE LINES

The battered Boromian ship landed at the transport facility on the surface of Yord, attracting a lot of attention. Shanna, Celine and Mariel waited anxiously as the hatch opened. Finally, the men stepped out of the ship, accompanied by Asheroff.

Shanna ran to Will and hugged him. "I was so worried about you!"

Will squeezed her tightly against him and replied, "I missed you."

Jack interrupted, "Where's Maya?"

"They're headed to Earth to reclaim Fleet headquarters," Shanna answered.

Will introduced the Boromian. "This is Commander Asheroff, our new ally."

"I never thought I'd meet a Boromian on friendly terms, Commander," Celine remarked, baffled by the turn of events.

"Nor did I think I'd ever call a human my ally," he responded.

"Mariel, please find Rethus and Atilena. We need their help," Will requested.

"What for?"

Asheroff interjected, "Because we could be on the brink of a war the size that none of us can imagine." Mariel quickly left them.

"Where's Bastille?" Will questioned Celine.

"He's helping Mynx and Neva with the vaults."

"Did they get the locks open?"

"Not yet, but Bastille had an idea."

"See how they're making out. I don't want to interrupt them if they're onto something important."

"You're briefing in the main hall?" Celine asked.

"Yeah." Will noticed somberness in Jack's behavior. "Everything okay, Jack?"

"Yeah. I'd like to get hold of Maya, though, if there's time."

"I understand. Do what you gotta do."

Will, Shanna and Asheroff walked together toward the main hall.

"You don't like us very much, do you?" Asheroff asked.

Will explained, "My mother was killed by your troops. My father and stepmother are missing and presumed dead at the hands of your people. It's difficult to put that behind me."

"It was you who attacked our armada in the solar system, wasn't it?"

"Yes, it was."

"That was very clever. I applaud you for your strategy."

"Thanks, but I'm not proud of killing anyone, even Boromians."

"Who were your parents?"

"My mother was Queen Seneca of Firenghia. She died fighting your soldiers."

"Ah, she was the shape-shifter."

"Yes she was."

"I was there when she fought my troops. We took many casualties from her. You should be proud of her."

"I am."

"What of your father?"

"My father and stepmother were murdered by your troops on Aramis – 5 a short while ago."

"Aramis – 5. Not a very good place to land."

"Did you have to kill them?"

"We didn't kill them. I witnessed a Fleet ship fire on them. Their ship took multiple hits and landed hard. The Fleet ship landed near them. That was all we saw."

Will's face reddened. "So you're saying that the Fleet is responsible for their deaths! I'm supposed to believe that?"

"When we kill, we kill for honor. I swear to you; we did not partake in this incident."

"Then I owe you an apology. I'm sorry."

"Maybe we can find common ground and leave our feelings of dislike behind us."

"We can try. Friends have to earn friendship and trust. Perhaps this is a good starting point. "

Asheroff explained, "There's one thing about Boromians you should know. We may be barbaric, but when we say something, we mean it."

"I hope so. We're going to need each other and the other races to fight together."

"This could be the battle of all battles in terms of greatness," suggested Asheroff.

"No battle is great, although it would be monumental in regard to the number of races fighting together for a common goal."

They arrived in the main meeting room and sat together at one of the tables. Shanna sat next to Will and held his arm. She leaned her head against him. Asheroff noticed her attire and commented, "You are dressed like a queen now."

"I am Shanna, Queen of Yord."

"And I am the King of Yord," Will added.

"That's quite interesting – from Fleet Officer to King."

Will replied, "Two lifestyles."

"When this is done, I hope you don't need two."

"I hope so, too."

Jack entered the room and sat with them. "I talked to Maya briefly. They are about to storm Fleet Headquarters."

"That's great!" Will exclaimed.

"I warned her that the Weevil are behind everything."

"Did you mention our new friends?"

"Yeah. She was... surprised."

"I figured she would be," Will remarked, convinced that he amazed her once again with his tact.

Atilena and Rethus entered the room. They were taken aback to see Asheroff sitting with Will and Jack. Rethus adjusted his electronic voice box and asked, "Am I missing something?"

Will greeted them, "Emperor Rethus and Empress Atilena, thank you for joining us."

"I never thought I'd see a Boromian sitting with a human, peacefully," Rethus remarked.

"I could say the same thing, seeing Urthonians as allies of humans," Asheroff admitted.

"Will has proved to be truthful in his quest for peace."

Will explained, "The Boromians and Attradeans have depleted their fighting forces. The Weevil have turned on the Boromians and will soon attack the rest of us."

Atilena inquired, "But aren't the Weevil subservient to the Boromians and the Attradeans?"

"We thought so but apparently that's not the case," Will replied.

Asheroff informed them, "They were never subservient to us, although we believed there was a co-existence between us. I am confident the Weevil have a very large force waiting to take over this quadrant."

"They wouldn't risk such an assault when the Attradeans and Boromians were strong so they used the two forces to help wear down the Fleet." Will explained. "To make that happen, they turned the two against each other."

"What is it that you propose we do?" Rethus inquired.

"First, we need to figure out what our fighting capacity is."

"I can offer forty-two cruisers and thirty fighters."

"Thank you, Emperor. We should have at least seventeen of the Fleet's battle cruisers joining us as well."

"Wasn't the Fleet taken over by the Weevil?" Atilena asked.

Shanna replied, "They were, but one of our people developed a weapon that exposes the Weevil imposters and renders them defenseless for a short period."

Rethus and Atilena were amazed. "That is excellent," replied Rethus. "The Fleet was a stabilizing force in the quadrant for a long time. If your people can purge the Weevil and regain control of the Fleet, then we can give the Weevil a battle."

Jack questioned Rethus and Asheroff. "Can either of you find out where the Weevil force is hiding?"

"Not if they're out of our scanners' range," Asheroff answered.

"We'll go now and make preparations. Tell us when you are ready to move on the Weevil." Rethus requested. Will and Shanna thanked the Emperor and Empress.

"What is your plan for confronting the Weevil?" Asheroff inquired.

"I'm not sure yet," Will said. "Do you have any suggestions?"

"I do. What if Tenemon and I take our cruisers to pursue the Weevil?"

"Why would you do that if there is a whole armada hiding out there?" Shanna asked.

"Because we can draw them into an ambush. You've been successful in ambushing the Attradeans and the Boromians already. Surely, you can come up with something for the Weevil."

"I can speak to Siphra about it. She's calling the shots now for the Attradeans," Will revealed.

"Very well. I will be on board my ship if you need to contact me."

"Thank you, Asheroff. We'll help you with repairs to your ship in the meantime." Asheroff thanked them and left the hall.

As Will pondered over the problem, Jack interrupted him, "I have an idea, Will."

"Go ahead. I'm listening."

"We use the *Leviathan* in cloaked mode. When the Fleet engages the Weevil head on, we strike from behind. We can use Tenemon's ship and Asheroff's ship to lure some of them from the battle, right into our sights."

"Perhaps. Then we can still keep the *Leviathan* a secret from the Weevil." Shanna interjected, "Something's not right about this, Will."

"What do you mean?"

"I think we're being lured into a trap," she commented.

"Let's see what Rethus comes up with first," Will decided.

"But what about Siphra and Tenemon?" Shanna asked.

"What about them?"

"Do you think Tenemon will roll over and let Siphra take power?"

Jack conceded, "She may have a point, Will."

"Let me think on this. We can't do anything just yet."

Shanna suggested, "Why don't the two of you get some rest? You've had a rough time of it."

"I could use some sleep," Jack admitted. Will agreed.

Shanna took Will by the arm and led him from the hall. "Shall we go to our room?" she asked.

Will frowned and said, "No. Let's go down to the *Phantom*. I need some quiet time away from all this."

"Should I leave you alone?"

Will pulled her toward him and startled her. "Why would you do that?"

"You need time to think, don't you?"

"I'll think a lot better after our honeymoon."

Shanna beamed at him and cooed, "Oh, Will."

They left the palace and descended through the secret walkway to the base. There, they boarded the *Phantom* and entered Will's quarters.

After spending some intimate time together, Will commented, "You know, Shanna, you might have a point."

"About what?"

"What if the Weevil expect us to mount an attack against them in some distant part of the galaxy?"

"Then we'd be vulnerable here on Yord," she responded.

"Not only Yord, but the entire quadrant. Now, how would we know for sure?"

"Simplify the whole situation into priorities. Start with the highest priority first."

Will laughed at her and teased, "Since when did you become a psychologist?"

"Arasthmus taught me," she replied proudly.

"And there is no one better to learn from."

"He really helped me a lot when you and Jack were gone."

"He's a good man." Will hugged Shanna then stood up and paced about the room. "Let's see. Our first priority is Maya and Steele's mission."

"What do we do if they succeed?"

"We have the Fleet on our side with Steele at the helm."

"What if they are defeated?"

"Then we only have Rethus' ships for strength."

"And what about the *Leviathan*? Where does it fit into our plans?"

"I don't want to advertise her or the underground base, at least not yet."

"Who can we trust among the real players in the group?" she asked, concerned by the growing number of former enemies on their side now.

"Well, I think we can count on Rethus and Atilena."

"I agree. What about Tenemon and Siphra?"

"Siphra will need our help. Tenemon is going to cling to his alliance, even if he isn't in charge."

"So Tenemon can be counted on to betray us?"

"Absolutely. I don't think he'll ever get over what I've done to him."

"What about Asheroff and the Boromians?"

"The Weevil killed some of his crew. I think Asheroff is thirsty for revenge."

"So why not send Tenemon's ship and maybe Asheroff's into the hornets' nest and see what happens? Keep your true allies close."

"I see you're becoming a strategist like me."

"Thank you," Shanna replied, proud that Will recognized her progress.

"But what about the Weevil?" Will conjectured. "Where would they likely hide a formidable fighting force?"

Shanna contemplated, "Could they be right in front of us and we just don't know it? Maybe they aren't in a distant part of the galaxy."

"Good point. Maybe they're using the Fleet as a front?"

"Maybe, but that's too obvious, isn't it?"

Will suddenly recalled the Weevil's interrogation on the Boromian ship. "You know, Shanna, when me and Jack were held prisoner on the Boromian ship, the Weevil kept asking us where Steele is."

"What's so important about Steele, especially since they already had control of the Fleet?"

"Maybe it's what they think she knows."

"Like what?" she asked.

"Think about it," Will replied. "What could she know that would be a threat to the Weevil?"

"I don't know. Maybe she could destroy them with a weapon or virus?"

"No. Think back. Who was interested in bidding for the *Leviathan*?"

"Tenemon, the Fleet and GSS."

"Exactly. The Leviathan-class destroyers would have given the Weevil supreme dominance of this quadrant even with the Attradeans and the

Boromians at full strength. Just one ship would make a decisive edge in battle."

"An edge for whom?"

"GSS. I'll bet Steele knows something about GSS that could blow up their plans."

"It was interesting that GSS would even bid on the *Leviathan*."

"Since Tenemon and the Fleet are out of the equation, maybe that's where we'll find the answer."

"Maybe Mynx and Neva can help with this," suggested Shanna.

"Let's pay them a visit."

When they entered the cargo bay, Bastille, Neva and Mynx were huddled around one of the vaults obtained from their earlier raid on the Ceratoan supply depot.

Bastille connected three wires to a circuit card inside the vault's locking device. Mynx and Neva watched anxiously. He looked up and commented, "Well, hello, royal lovebirds. Sorry I missed all the festivities. I'm in real deep with this stuff."

"No problem. I can't believe you duplicated the cloaking device on the *Leviathan*," Will said.

"It's better than duplicated. Even the energy trails can't be traced," Bastille said, proudly.

"Did you find a way to cloak torpedoes?"

Bastille chuckled and asked slyly, "Why?"

"You are slick, Bastille," Will responded, knowing that he had accomplished that feat.

"So, what is going on with the vaults?" Shanna asked.

"Bastille has an idea to open the locking mechanism," Mynx explained.

"Do you think it's really worth all this trouble?"

Neva answered, "These vaults are only used for very important information and for valuable objects. I'm sure there's something worthwhile inside."

"If you say so. I'm just asking," Shanna remarked, unconvinced of its value.

"I'll be ready in a minute. Be patient with me," Bastille said.

"Take your time. I really wanted to speak with Mynx and Neva," Will told him.

"What about?" Mynx asked.

"GSS. When Jack and I were captive, the Weevil were quite concerned about Steele's whereabouts. Think back to when we captured the *Leviathan*. Why would GSS bid on it? I think there's a connection somewhere."

"What kind of a connection are you looking for?" Neva asked.

Shanna explained, "We think Steele knows something very important that could link GSS and the Weevil threat."

"I don't know of anything obvious that would link the two," Mynx answered.

Neva pondered for a moment. "Maybe Steele doesn't realize she knows something that threatens them."

Bastille added, "She was very interested in the contents of these vaults. She said she couldn't wait any longer and bolted out of here." He pressed a switch and the locking device clicked open. "Voila! I told you it would work."

Mynx approached the vault cautiously and opened it. Everyone waited anxiously to see the contents.

The vault had three shelves in it. On the top shelf was a packet with four disks. The packet was marked, "Tarsus: Review and Destroy" The second shelf had a document listing the Gallian military chain of command. On the bottom shelf was a thick, leather photo album. The cover title read, "Primary Targets – Terminate ASAP."

Mynx took out the photo album and set it on the table. Everyone stood by and watched as she turned the pages. The first four pages meant nothing to them. The fifth page shocked Will. There were photos of his father and stepmother. Each photo had a red diagonal stripe across it.

"What the hell is this?" Will bellowed angrily.

"What's wrong?" Shanna asked.

"That's my father and my stepmother Tera."

Neva looked at the photos. "Now, I guess you know who was behind their deaths."

Will uttered, "So Asheroff told the truth. Those bastards!"

Mynx turned more pages. She stopped and stared at a picture of her and one of Neva. Under each picture was a comment: "Terminate with extreme prejudice."

Will quipped, "Looks like you made the hit parade as well."

Neva kidded, "I guess they didn't like what I did to Nestor."

Mynx was visibly shaken. "Screw those parasites!" she shouted,

Will put a hand on her shoulder. "Easy, girl."

"I'll show them extreme prejudice," she shouted. Neva laughed at her.

"What's so funny?"

"I never thought I'd see you act like me."

"Sister, you ain't seen nothing yet! I only need one arm to take care of those scumbags."

Will reminded her, "Don't forget, Mynx, we got this far by being smart."

"Yeah, but now it's personal." Mynx turned the page. She chuckled sadistically as she saw pictures of Will and Shanna. "Well, lookie here."

Will frowned at the picture and complained, "You think they'd at least get my good side."

Shanna studied the pictures briefly and said, "Those shots were taken at Eve's. They replaced the background with white."

"How can you tell?"

"First, I remember wearing those clothes and second, I wore eye shadow and mascara that night."

"Is that the only night you wore eye shadow and mascara?" Will asked.

"Yes. I wanted to attract other men and make you jealous."

"Ah, that's right. You had Weevil hanging all over you," he teased.

Shanna was embarrassed. "Can it, Will."

Bastille asked, "Do you think the Gallians could be in league with the Weevil?"

"Maybe we should find the Gallians and ask them," Will suggested.

Mynx stared at Will. "Do you even know who the Gallians are?"

"Sure. We're best friends," he kidded.

"I wouldn't be surprised."

"No, I'm joking. I will find out, though."

Bastille moved his equipment to the second vault. Shanna asked him, "How long does it take to open one of these?"

"A little while. The hardest part is removing the cover plate without detonating the inside of the vault."

"Detonating! Like in a bomb?"

"Absolutely."

"I'm sorry I asked."

Mynx took the packet with the disks from the vault. "I'll run these through the computer and see what's on them."

"If it's anything vital, let me know right away. Meanwhile, I'll find out who or what the Gallians are," Will said.

Bastille shook Will's hand. "By the way, congratulations to you and your lovely Queen."

"Thank you, Bastille."

Bastille stood and hugged Shanna. She blushed and thanked him.

Will remarked proudly, "It's great working with the best team in the universe."

Bastille answered back, "It's great to be a member of the best team in the universe." They enjoyed a good laugh together.

Will and Shanna left the *Phantom* and sat on a bench near the steps. "What now, my king?" she inquired.

"There's a lot going on. We have to sort this out into a logical plan of action."

"Maybe we should sleep on it," she suggested, smiling coyly.

"For some reason, I knew you'd say that."

"After all, Will, it is our honeymoon."

"Ah, yes. Perhaps some red wine and a little privacy will do."

Shanna took Will by the hand and led him up the stairs. They returned to their new bedroom in the upper level of the base.

"We've got to narrow this down to one bedroom," complained Will. "It's too confusing."

Shanna ribbed him, "Why? Each one is filled with fun memories."

Will kissed her cheek and pulled her close to him. "The memories are fine. I'd just like to have one that I can call home."

· · · · · · ● ● ● ● ● ● ● ● ● · · · · · · · ·

Later that evening, Mynx sat at the computer in the cargo bay. She inserted the first mini-disk into the drive and waited for the monitor to activate. Bastille was nearby and continued to work on the second vault. He lifted the cover plate from the locking mechanism and slid it back. Suddenly, the vault emitted a bright flash, which filled the room. He screamed, covered his eyes, and shook violently. "My eyes! My eyes!" he cried.

Mynx panicked. She rushed over and quickly helped him into a chair. "Don't move. I'll get help."

Still covering his eyes, Bastille cried, "Hurry! They're burning!"

Mynx rushed to the intercom and shouted, "Arasthmus, report to the *Phantom*'s cargo bay - ASAP!"

Will and Shanna heard the page and also hurried to the cargo bay. When they arrived, they saw Bastille in distress. "What happened, Mynx?" Will shouted.

"The locking device exploded and the flash burned his eyes."

"I'll find Arasthmus," Shanna said.

"I paged him on the intercom," Mynx told her.

"He might be with Mariel in the hall. I'll go look." Shanna rushed from the *Phantom* and hurried across the deck to the granite stairwell.

Will opened the first-aid kit on the wall and removed a bottle of solution and two bandages. He poured the solution on each bandage.

"Bastille, I have two wet bandages. Let me cover your eyes with them."

Bastille cried, "I could be blind!"

"You don't know that." Will pushed Bastille's hands slowly away from his eyes then held a bandage on each of them. "Does that help?"

"A little." Bastille moved Will's hands away and held the bandages in place by himself.

Mynx was visibly upset. She sat on the chair in front of the computer and stared sympathetically at Bastille. Will looked at the second vault. The door had sprung open and smoke streamed from the locking mechanism. He peered inside and saw a corroded packet on one of the shelves. "Son of a bitch!"

"What's wrong?" Mynx asked.

Will removed the damaged packet from the vault. He examined bits of writing on the cover. "It's from Earth. Unfortunately, we'll never know what was in it."

Bastille sobbed, "Don't tell me I did this for nothing."

"No. I think we can presume that the Weevil have infiltrated Earth's defenses as well."

Shanna and Arasthmus entered the *Phantom* and descended the stairs leading into the cargo bay. Arasthmus knelt in front of Bastille and gently removed the bandages. Bastille's eyes were swollen and the pupils were pale colored. Bastille asked tearfully, "Is the damage permanent?"

Frowning, Arasthmus replied, "I don't know. We'll give it a few days for the swelling to dissipate, then we'll see if there's improvement."

"What if it doesn't?"

"Then you're permanently blind," Arasthmus answered somberly. Bastille buried his face in his hands and cried.

"You should lie down. I have some salve that might help," Arasthmus suggested as he escorted Bastille up to the nearest cabin and helped him lay down on the bed.

Will pulled up a chair and sat next to Mynx. "This really sucks."

Mynx agreed. "I couldn't believe it. We were just talking and then there was a bright flash."

Will noticed the computer monitor. "What are you looking at?"

Mynx glanced at the monitor. "I don't know."

They suddenly realized that they were viewing a map of worlds in seven quadrants conquered by the Weevil and the dates that their conquest occurred. "What in the world, Mynxie?"

She studied the map and mumbled aloud, "Well, I'll be..."

"Look at how many worlds they've taken over!"

"They're like parasites that ravage everything they touch."

"Yeah, but these parasites are intelligent," noted Will.

Mynx scrolled through several other menus. She noticed a small report on Gallia. "Maybe this will help answer your questions about the Gallians."

She clicked on the planet Gallia and watched in awe as a large report from the Weevil high command appeared on the monitor. They gazed at the monitor and read through the report.

"It appears Gallia has critical ores beneath its surface," Mynx announced. "Most of the report is about it."

"Why would the Weevil be so concerned about ores?" Will asked.

"Who knows?"

"I think we need to visit the Gallians. Maybe the Weevil were unsuccessful in conquering them and they're unhappy about it. Perhaps we can learn something from them." Arasthmus returned to the cargo bay, capturing their attention. "How is Bastille?" Will asked anxiously.

"Not good. I gave him something to make him sleep."

"Do you think he'll recover?"

"I can't tell. I've never dealt with eye injuries before."

"Thanks, Arasthmus. I know you're doing your best." Dejectedly, Arasthmus left them.

As Will stared at the monitor and pondered, Shanna suggested to him, "Maybe there aren't really Gallians on that planet."

"What are you getting at, Shanna?"

"Maybe somebody from another world wants to stop the Weevil from taking the ore."

"I don't get it. What makes you think that?"

"Just a guess. It's strange that the Weevil conquered everyone else with ease but couldn't conquer a single planet like Gallia."

Will agreed. "It's possible. There must be a clue out there."

"Do you think it's worth pursuing Gallia over this device?" Mynx asked. "Maybe we should be looking at other options."

"I have a strong hunch we'll find more than we bargained for on Gallia."

Regent and Kiera entered the main quarters. Will stood and shook Regent's hand. "Well, hello, strangers," he said. "Congratulations on your wedding."

"Thanks. I'm sorry we missed yours," Regent said.

"I guess you heard about Bastille."

"Yeah. Kiera overheard Shanna talking with Arasthmus. Will he be alright?"

"We don't know."

"Will, I have a request."

"What is it?"

"We've spoken with Queen Siphra about Attrades. I'd like to lead a recovery team there." Will leaned back against the table and stared at Regent, curious as to why he would want this.

Sensing his concern, Kiera explained, "They are going to recolonize their planet from scratch. During the interim, many of the survivors will need a place to stay. We can bring them here."

"It would be a humane gesture on our part," Will remarked.

Regent added, "I know this is a surprise to you but we can handle it in relatively quick fashion."

"I know you can. I'm weighing out the possibility of things going wrong down there, especially with Tenemon."

"Breel and Laneia want to accompany the queen as her body guards for that very reason."

"Take the *Phantom*. Saphoro will be your pilot."

"Where is she?" Regent asked. "I haven't seen her in quite some time."

"She's working with Kahlin and Talia on board the *Leviathan*. Let her know it's urgent that she help with this," Will instructed.

Regent extended his hand to Will. "Thank you. We'll do you proud."

"I know you will."

Regent and Kiera departed the *Phantom*. Will sighed and followed them.

·········●●●●●●●●●●·····

Inside Fleet headquarters, Steele and Maya led a small contingency of Fleet officers. They fired several shots down the hall at three resisting officers. One returned fired. The round grazed Steele's cheek. A small stream of blood trickled from the wound. She wiped the blood off and shouted, "That son of a bitch! Cover me."

Maya fired a steady barrage of pulse fire at their quarry. Steele crawled quickly down the hall. She alertly pointed her pistol in the direction of the Weevil imposters.

One of the officers stepped from behind the corner to fire at Maya. Steele targeted his head and ordered, "Don't move or you're dead." The officer dropped his weapon and put his hands up.

Steele then ordered the other two, "Drop your weapons and come out now!" They stepped into sight. Maya approached the prisoners and held them at gunpoint. Steele stood up and casually brushed herself off.

One of the captives yelled at her, "What's wrong with you? You sold us out, didn't you?"

Steele grabbed the man's chin tightly and glared at him. She stated coldly, "We're going to find out who sold out whom, Solo." She reached into her shirt pocket, took out the small spray bottle she received from Arasthmus and pointed it at him.

"This is ridiculous," complained the officer. "I protest."

She sprayed Solo in the face. He gagged and then fell to the ground, convulsing.

"Just as I thought," Steele uttered. She aimed her pistol and fired a shot. The burst of energy obliterated his head. His body quivered until the flesh fell off, revealing a Weevil body. The other officers stared in disbelief. Steele sprayed each of them in the face before they could react. One of the men sneezed and wiped his nose.

The other appeared to hold his breath for a moment. After several seconds, his face turned pale and he too gagged, fell to the floor and convulsed. Steele fired a shot at his head, which exploded and left a quivering corpse. The flesh and clothes fell away, revealing another Weevil. The remaining officer looked ill. He trembled in fear.

Steele taunted, "I guess the Weevil didn't see you as a threat, Manderville."

"Don't kill me, Commander," he begged. "I didn't do anything wrong."

"Now that's the coward I know."

"But how did you know I wasn't one of them?"

She sprayed his face again and goaded him, "Quiver and you're dead." Manderville feinted on the floor.

Maya grabbed him by the collar. "Get up!" she ordered. He came to and clumsily got to his feet. Maya picked up his pistol and handed it to him.

Steele explained, "We're taking back the Fleet from the Weevil. Are you with us?"

He looked relieved and answered, "Of course, I am."

"Good. Let's go."

"That spray really works, huh?" Maya asked.

"Sure does. Now, we can kick their Weevil asses right out of here."

· · · · · · ● · · · · · · · · ·

On the second level of the temple, Will, Shanna and Arasthmus stood by Bastille and observed him as he slept. "Boy you really knocked him out, Arasthmus," Will commented.

"He's got to rest his eyes if there's going to be any chance of recovery."

"Thanks again for your help. I don't know what we'd do without you."

"This is the tough part of being a doctor – seeing your friends injured and hoping you can help them recover."

"You're right about that. I guess we'll check on Mynx and see what else she's found on that disk." Will and Shanna descended the stairs to the main floor.

Arasthmus waited until they were out of sight and returned to the upper level of the temple. He went to Mariel's door and rapped lightly. Mariel opened the door and looked both ways. The hall was empty, except for Arasthmus. She pulled him inside her room and closed the door.

Shanna paused on the dock and smiled. Will asked, "What's wrong?"

"Arasthmus is up to something. He was in an awful hurry to leave here."

"What are you thinking?" he inquired, knowing Shanna's devious nature.

"I think he and Mariel have something going on."

Will chuckled and commented, "Wouldn't that be cool?" They left the *Phantom* and hurried to the upper-level hallway.

"Where did he go?" Shanna asked.

"I wonder if he's in her room," kidded Will.

Shanna's eyes widened in surprise and her jaw dropped. She whispered excitedly, "Let's find out!"

Will was embarrassed for even considering it. "I don't think that's a good idea, Shanna."

"Come on. I owe Mariel for all the times she rudely walked in on us."

Will replied uneasily, "I think I'll wait right here."

Shanna tiptoed to Mariel's door and paused. She listened carefully for a moment, then looked back at Will and affirmed with a nod that Arasthmus was inside. She knocked softly on the door then opened it.

"Mariel, I wanted to... Oh, my gosh."

Mariel and Arasthmus were cuddled in each other's arms in her bed. Their bare shoulders uncovered by the white sheet revealed more than enough detail of their actions. "You shouldn't ever enter someone's room unless you're invited!" Mariel shouted.

"I'm sorry. I didn't know," Shanna said giddily.

"Oh, dear," Arasthmus muttered.

"Leave, Shanna!" Mariel bellowed.

"Okay. We'll talk later, Mariel."

"Now!"

Shanna closed the door and covered her mouth to hide her laughter. She and Will raced down the stairs playfully.

Will asked, "Well, were they?"

Shanna said giddily, "Oh, yeah!"

"Oh, yeah, what?"

"Oh, yeah, they are definitely involved and I don't think this just started."

"Well, I'll be. I'm proud of Arasthmus."

"Wait until I see Mariel. I'm going to lecture her about morals. She was in a very compromising position for a woman."

"Mariel's done a lot for us. I think you should cut her some slack."

"Oh, all right."

"Let's go to the hall for a drink. I've got to figure out a plan for Gallia."

• • • • • • • • • ● • • • • • • • • • •

Steele and Maya led twenty-five Fleet soldiers to the transport bay. The hatch on one of the ships closed and sealed. Steele ordered the nearest soldier, "Cut power to the umbilicals on that craft as well as the gates. That ship cannot leave." The soldier obediently rushed to the control power console.

Steele declared proudly, "This should be the last of the Weevil on this base."

"They didn't put up half the fight I thought they would," Maya remarked.

"I don't think they ever expected us to come after them like this."

Steele put on a headset and plugged the comm cord into a port on the side of the ship. "Surrender and we'll spare your lives," she warned. "If not, you'll die like the rest. This is your only chance."

The ship's hatch opened and twelve Fleet Officers emerged, hands on their heads.

"Imperial General, this is a mistake," one officer said.

Steele approached the man and replied sarcastically, "You really think so." She sprayed him in the face and watched patiently.

The man coughed and gagged violently. He fell to the ground and shook until his flesh and clothing fell away from his body. Steele aimed her pistol and fired. The Weevil's head exploded, sending pieces of black and green flesh across the deck.

"Anyone else want to dispute their condition?" None spoke.

Steele ordered, "Twelve of you escort the prisoners to the brig. The rest of you search that ship. I believe we're missing three." The soldiers cautiously entered the ship.

Steele thanked Maya and declared proudly, "The Fleet is ours again."

"I'm proud to be part of this. What's next for us?"

Steele deliberated for a moment and answered, "I guess you'll be returning to your friends. I'll have to stay and reorganize our forces for a state of readiness."

"So you are Imperial General Furey again?"

"Yes, I'm afraid so. It was fun, though, working with your space pirates for a short time."

"It was a pleasure working with you, Steele."

"Thanks for everything, Maya. Keep up the good work."

Maya saluted Steele. "Good luck, Imperial General."

Steele returned the salute. "You, too, Maya. I'll be in touch soon."

Maya departed the transport bay, feeling elated over her part in the successful retaking of Fleet Command.

· · · · · · ●●●●● ● ●●●●● · · · · · ·

As the *Phantom* landed just outside the ruins of Tenemon's palace, Tenemon and Mantos watched from a distance. "It looks like help has finally come!" Mantos exclaimed.

Tenemon bellowed in disgust, "Yeah, in my ship."

"Perhaps, considering our predicament, we should let bygones be bygones." Mantos suggested.

Tenemon angrily grabbed Mantos by the arm and bent it behind him. He placed his pistol against the side of Mantos's head. "Let's not forget who is king around here, huh, Mantos?"

Mantos quivered and replied, "My mistake, sire. I didn't mean anything by it."

"Neither did I. Watch yourself."

A number of Attradean soldiers and civilians approached the *Phantom*.

Queen Siphra was the first to exit. She was saddened by the degree of destruction to the cities on her world. Laneia and Breel came out behind her, carrying food and water. They dispersed the parcels to the Attradean survivors. Regent and Kiera emerged from the ship carrying medical supplies. They set up a station just outside the hatch.

Tenemon and Mantos appeared and passed through the group. Siphra immediately spotted him and cornered him.

"It took you long enough to get here," Tenemon complained.

"No thanks to you. It's difficult making allies with races that you attempted to annihilate."

"That's the responsibility of a king."

"Well, you won't have that responsibility anymore. You are officially relieved under the Attradean code of government."

Breel and Laneia eyed him, knowing he would not go quietly. Tenemon smiled and looked around at the people near him. He drew a pistol from under his armor and fired at Siphra.

Breel saw Tenemon draw the gun and dove in front of Siphra. The shot struck him in the chest and he fell to the ground dead. Laneia drew her pistol and fired at Tenemon, wounding him in the shoulder. Tenemon fell to the ground. He raised his arm to fire at Laneia but she fired again, striking his arm. His weapon fell harmlessly away from him. He moaned at the pain in his badly injured arm. "Damn you, Laneia!"

She stepped on his chest and reminded him, "I told you the next time we meet, I would kill you."

Tenemon challenged her, "You don't have the gall."

Laneia fired a shot into his leg. "How does that feel?" She fired another at his knee. Tenemon screamed in pain.

Siphra watched stoically as Laneia methodically fired at various other parts of Tenemon's body. Tenemon lay in a bloody pool of green blood. He muttered, "Kill me, you bitch."

Laneia looked at her mother, Siphra, and then back at Tenemon. Siphra was content to let Laneia handle Tenemon's disposal.

Laneia ordered Mantos, "Hang him from his ankles. Prepare him for the chordites."

"No! Not the chordites," Tenemon screamed. "Just kill me!"

Laneia looked at Breel's lifeless body. She took a knife from his belt and knelt by Tenemon. Four soldiers held Tenemon down against the ground. She grabbed his tongue and stretched it out.

Tenemon groaned as he tried in vain to pull away. Laneia cut the tongue off and held it up for everyone to see. "Long live Siphra, Queen of Attrades," she shouted.

The survivors cheered and chanted, "Long live Siphra."

Laneia knelt by Breel and sobbed. Siphra knelt next to her and consoled her. The Attradean soldiers tied Tenemon's legs together and hung him from a tree at the edge of the forest. They walked away from him and rejoined the crowd. The rope soon turned white as leech-like insects slid down to Tenemon's bloody body. They grew in number until they covered him.

Laneia and Siphra watched with satisfaction as Tenemon's wriggling body was soon encased in a cocoon. The last thing they saw were his eyes, wide in fear and in agony, before they were covered by the leeches. Green blood dripped from the cocoon and formed a pool on the ground.

Siphra announced impassively to those around her, "Justice has been served."

· · · · · · · · · · ● · · · · · · · · · · ·

Will and Shanna sat at the head table in the hall. Will stared at his glass of wine and swirled it around. Shanna leaned her head against his. "You're really worried about Gallia, aren't you?"

"Yes, I am. I sense something very important is going on there, but I also sense that we'll be in grave danger."

"But, Will, everything we do puts us in danger."

"I know. It's just that there's something about this I don't like. This is different."

Mariel entered the hall and sat down with them. She poured herself a glass of wine and stared at Shanna.

Shanna leaned forward and kissed Mariel's cheek. "I love you, Mariel. You're the best mom in the world."

Mariel blushed and sipped her wine. "But I'm not your mother," she replied sadly, wishing that she had a family of her own.

"It doesn't matter. If she were alive, she'd be just like you."

"Look, Shanna, about before…"

Shanna laughed. "I'm so glad for you. I think you needed that."

"I was going to tell you that if you ever enter my room like that again, I'll…"

Shanna cut her off, "Don't worry. I just wanted to get you back once for all the times you interrupted us." Will blushed and covered his face.

Mariel sighed. "Alright, Shanna. You win."

"Can I ask you one question, Mariel?"

"Yes."

"How long have you and Arasthmus been seeing each other?"

"Shanna! That's none of your business!"

"How does Regent feel about you stealing his chess partner?"

"Arasthmus can play chess whenever he chooses."

"I'm glad for the two of you. I hope things work out," Will told her.

"Thank you, Will."

"We're going to Gallia," Shanna informed her.

"Gallia! For what?"

Will explained, "I'm not sure but I think we'll find the answer to the Weevil incursion there."

"Do you know something about Gallia that we don't?" Shanna asked Mariel.

"I know it's an evil place. A long time ago, there were stories about merchants who landed there to collect valuable minerals and ores. They never returned. The abandoned ships can be seen on the planet's surface, but no one dares to go there to find out what happened."

Will stared at his glass and nervously swirled his wine. "Are you sure we want to do this, Will?" Shanna asked.

"Yes, I am. I'm thinking that maybe you should stay here."

"No way! I'm going with you."

Mariel grew more concerned about them. She inquired somberly, "What if something happens to the two of you? Who'll rule Yord?"

"This could be bigger than ruling Yord," Will told her.

"Then you must do what you think is right. I'll leave you now to consider your options." She finished her wine and set the glass on the table. "Please be careful. You both mean a lot to me."

"Even me?" Shanna asked.

"Yes, you little brat. Even you." Mariel smiled at her and left the room.

Mynx entered and sat with them. "You don't look so good, Will."

"I'm fine. I'm just lost in thought."

"Did you find out anything new?" Shanna questioned Mynx.

"As a matter of fact I did. First of all, Maya's on the way back. Steele is running Fleet Command again."

"Excellent," Will replied.

"I've been through the disks and analyzed some of the other information from the vaults. It appears that each vault was sent from a different location to be picked up by a third Weevil party. The first vault came from GSS. The second came from Earth."

Will became attentive. "GSS and Earth, hmm."

"That reinforces our theory that GSS is a front for the Weevil," Shanna responded.

"Exactly, but Earth, too? That's very interesting,"

"The Fleet moved its headquarters to Earth before your encounter with the Boromians," Shanna pointed out. "Could that have something to do with it?"

"I wonder if they're looking to use the Fleet's portals," Will pondered. "Only certain personnel had access to the location and codes, including me."

Mynx asked, "Why wouldn't the Weevil use their own portals?"

"Maybe they don't have any."

Shanna suggested, "Maybe that's why they attacked you at Eve's. Perhaps they knew you were an officer and would have that knowledge."

"Maybe. Not all officers had that information, though."

"Could someone have told them?"

"Like who?"

"I don't know."

"Every day we have more questions than answers," Mynx complained.

"The one thing we know for sure is that the Weevil are planning to conquer everything and everybody," Will stated. "Our immediate mission is to thwart their invasion. Later, we'll pursue them into their own territory."

"How big is their invasion force?" Mynx asked.

"According to Asheroff, it's huge."

"Can you trust Asheroff?" Shanna asked.

"He has no reason to lie to us," Will said.

"Does Earth have any resources to fight the Weevil?" Mynx inquired.

"I'm not sure. I know very little about them," Will told her.

"Maybe you should investigate the situation on Earth first before the Weevil take over."

"No, I think Gallia comes first."

Jack rushed into the hall. "Well, well. I'm glad I finally found you, Will."

"What's up, Jack?"

"I spoke with Maya. She'll be here soon."

"And..."

"Laneia, Regent and Kiera are staying on Attrades to help out with the recovery."

"What about Breel? I could use his help."

Jack informed him somberly, "Breel's dead."

"Tenemon?"

"Yeah."

"What a shame."

"Tenemon's gone, too."

"How?"

"Public execution. No details, although I did hear that Laneia warmed him up with several well-placed shots to his extremities."

"That poor girl. Tenemon was everything to her," Will kidded sarcastically.

"I have an idea. Let's get out of here," Mynx suggested to Shanna.

"Where are you going?" Will asked.

"Don't worry about it. We'll see you in a bit," Shanna said. The girls left the room.

Jack continued, "Maya said they've got control of Fleet headquarters again."

"So I've heard."

"No, this is the good news. Steele is back at her old job, running the Fleet."

"Really?"

"Yeah, isn't that great?"

"It just seems like it was too easy," Will remarked. "Besides, I wanted to question her about why the Weevil were so interested in her."

"Well, I'm still grateful that she's back in her saddle again. Besides, I'm sure the Weevil know where to find her now. Happy hunting!"

"Come on, Jack. She wasn't that bad."

"Maybe not, but she and I have a very unpleasant past."

"Well, it sounds like she's gone, so you can relax." Jack chuckled at his remark. "What's so funny?" asked Will.

"This is really sad. We need to get out more often."

"It's funny you should say that."

"Tell me we're going out for some R and R," Jack begged.

"We are. We have a date with destiny on Gallia."

"Gallia! What for?"

"Answers, I hope."

"They'd better be good ones. I don't know many people who came back from Gallia. Actually, I don't know any."

"When Maya returns, we'll make the trip. She and Shanna will join us."

"Do you think it's wise to take the girls to Gallia?" Jack asked.

"I've been thinking about that. I think we'll leave them on the ship until we see what's going on. We may need them to rescue us if we get into trouble."

"I'll buy that. When do you want to go?"

"How about morning? It'll give me time to brief the others on what's going on."

"Morning is fine."

"Then I'll see you at breakfast."

"Yeah, breakfast." Jack replied, concerned. He left the hall with a worried look on his face.

Will considered his dilemmas and sipped from his glass. "So many questions and so few answers," he repeated Mynx's words to himself.

Neva entered the hall and sat down across from him. "Hello, Will."

"Hi, Neva. How are you?"

"Not bad."

"Beat anyone to death lately."

"No. I miss it, though."

"What have you been up to?"

"Monitoring our friends at GSS. Deciphering coded messages. Boring stuff like that."

"Anything exciting?"

"Maybe. There's an awful lot of traffic emanating from under the GSS space station."

Will grew interested and sat up straight. "Don't stop now, Neva."

"I think the large, ringed structure under the station is a portal and the Weevil could be using it to cross over from another part of the universe. We assumed it was for positioning the station."

"How about the coded messages?"

"Haven't cracked them yet. Bastille was doing much better with them than either Mynx or myself. Unfortunately, he's not able to work on them right now with his eye injuries. I can't trace the source which leads me to believe it's coming from the other side of the portal."

"You mean they can communicate through the portal?"

"It appears that way."

"That's unheard of! What can we do about it?"

"I think we should board the station, kill everyone and blow it up."

Will was amused as usual by Neva's zest for death and destruction. "Thanks for your candor."

"Anytime. By the way, I'm serious. If the Weevil control GSS, there's nothing to salvage. Blow the bitch up."

"I'll consider that option. We need to know a little bit more about the space station as far as the number of personnel on it, before we think about boarding it."

"Perhaps I can get close enough to scan it. That'll get us some information."

"Now that we can do."

"How soon?" she asked.

"Let's plan for tomorrow. You, Mynx, Celine and Talia will take out the *Leviathan*."

"Now we're talking!" Neva shouted excitedly and high-fived him.

"Jack, Shanna and Maya will join me on the *Phantom*. We're going to Gallia. You're our muscle until I get back."

"It's about time we got some action! Thanks, Will."

VIII

GALLIA

Will awoke in the master bedroom of the palace with Shanna was nestled against his side. He slid away from her and sat up on the edge of the bed. He grilled himself with questions over Gallia and wondered, *What if it isn't the ore that the Weevil want? Maybe there's something else.*

He donned a robe and left the bedroom. As he walked down the hall, he heard giggling from Mariel's room. *That's funny,* he mused. *Just like Shanna, Mariel's like a little girl in love. Who would have thought?* He took two steps and sneezed. The giggling stopped. *Oh, geez. They're going to be mad at me,* he thought.

The door slowly opened and Arasthmus peered out. "Are you alright, Will?"

"Uh, yeah. Sorry, Arasthmus. I didn't mean to bother you."

"No problem. Where are you going?"

"Downstairs for something to drink. I can't sleep."

The door opened wider and Mariel's face appeared. "We can join you if you want to talk."

Will was apprehensive but Arasthmus suggested, "I could go for something to drink as well."

Mariel declared, "Then we'll join you. I'll make something for us to nibble on."

"Are you sure? I really didn't mean to bother you."

"Don't worry about it. We were just talking," Arasthmus explained.

They descended the stairs and entered the dining room. Will and Arasthmus sat down at the long, glass table. Mariel left them and went into the kitchen.

"What's bothering you, Will?" Arasthmus asked.

"I'm worried about Bastille. How's he doing?"

"Some of his vision has returned. We're hopeful that he'll recover more over time."

"That's good news."

"Celine has been with him the whole time."

"They're a unique couple," Will quipped.

Mariel returned carrying a silver plate with five sandwiches and a pitcher of hot tea. Will was surprised to see the tea. "I haven't had hot tea in ages."

"This is from the first crop of tea leaves we're growing on the surface," she announced proudly. "It's been many years since we attempted anything on the surface."

A servant brought napkins and three cups, set them on the table. Mariel thanked her and she departed the dining room.

"You're still going to Gallia?" Arasthmus inquired.

"Yes, in the morning."

"What in the world for?"

"There are answers there that we need to have."

"What kind of answers?"

"For starters, what role does Gallia play in the potential Weevil invasion?"

Arasthmus warned, "No one that I'm aware of has ever survived a visit to Gallia."

"Jack, Maya and Shanna will join me on the *Phantom*."

"And what are you going to do once you get there?"

"Jack and I will land on the surface. Shanna and Maya will monitor us from the ship in case we encounter resistance." They sipped their tea and ate sandwiches quietly.

Mariel finally broke the silence and said, "What if something happens to all of you?"

"Then, I think you and Arasthmus should rule Yord."

"That's not funny," she responded, frightened.

"No, it's not, but it is reality. This is a dangerous risk but it's got to be done. The Weevil are amassing for an invasion that we're not ready to counter."

"The Priestesses have seen this invasion through the Eye," Mariel said. "They could not define when or where, although it appears to be massive."

"I can ask Rethos if he's received any information on the location of the Weevil armada."

"What if you are being lured away from here for a reason, Will?" Arasthmus inquired.

"I considered that. The *Leviathan* will stand by, cloaked in a defensive position. If the Weevil try anything, our people will do their best to squash them."

"What about the baby? Is it smart to take Shanna with you?" Mariel asked.

"She won't have it any other way, but that's why I want her on board the *Phantom* with Maya."

Shanna entered the dining room and frowned at them. "Thanks for inviting me to the party," she said, annoyed to be left out.

"I couldn't sleep. Come sit by me." Will patted the chair next to him. Shanna pushed her rumpled hair back behind her shoulders and sat next to him. Will pulled her close to him. He held his cup of tea to her lips and she sipped from it.

"Thanks, baby."

Will blushed. "Do you have to call me 'baby'?"

"Yes, I do."

Mariel announced humbly, "Since you are both here, there's something we want to tell you."

Shanna kidded, "Don't tell me, Mariel, you're pregnant."

"Heavens no. Not at this point of my life." She hugged Arasthmus, gave a wide smile and announced, "Arasthmus has asked me to marry him and I've consented."

They were stunned by the news. Will shook Arasthmus' hand. "Congratulations to you both."

Shanna gazed at Mariel until tears streamed down her cheeks. She blubbered, "I'm so happy for you, Mariel." They embraced in an affectionate hug. Mariel also sobbed as she clung to Shanna.

"When are you planning to be wed?" Will asked.

"In four cycles," Arasthmus replied.

Mariel sniffled and pleaded, "Please don't get hurt on Gallia. We want you here with us for the wedding."

Will kidded, "Now we really have to be careful. I wouldn't miss this for anything." They finished the sandwiches and tea.

"Now," Arasthmus announced, "I'm ready to go back to bed."

"After hearing your good news, maybe I'll forget about Gallia and get some sleep, if Shanna will let me," Will mentioned playfully.

Arasthmus and Mariel stood. She urged them to be careful and to return safe. They left the dining room, holding hands.

Will smiled at Shanna. Her hair was a mess and her short robe was wrinkled. She wore fuzzy, pink slippers and looked tired. She groaned, "Don't stare at me like that. I look awful."

"No, you don't. You look fine."

Shanna leaned on his shoulder and hugged him. She stood up and turned sideways. "Do I look like I'm showing yet?"

Will studied her figure and replied, "Not yet. Perhaps I should look a little closer, though." He pulled her close, kissed her and slid his hands along her back.

Shanna remarked coyly, "That's not where a baby shows, you naughty boy."

"Let's go upstairs and I'll look more closely at the rest of you."

· · · · · · ●●● ● ●●● · · · · · ·

The next morning, they arrived at the transport bay, holding hands. Maya and Jack sat just outside the hatch of the *Phantom* on a concrete bench, conversing cheerfully. Will greeted them, "Good morning, friends."

Maya teased, "Should I thank you for giving Jack a reason to spend some time with me?"

Jack winked at them. "I told her you ordered me to spend the night with her."

"Oh, yes, Maya," Will laughed. "He was miserable without you."

Maya looked proudly at Jack. "He's like a little kid. He's got to have his candy or else."

"And did he get it?" Shanna teased.

"Of course. I'm surprised he doesn't have a tummy ache."

"Alright, that's enough, ladies," Jack uttered, his face red.

"Are you ready?" Will asked.

"I can't say I agree with this plan of yours, but yes, we're ready," Maya replied.

Will looked at Jack, who replied uncomfortably, "If Maya is, then so am I."

"Oh, please, Jack. Don't be a kiss up," Will admonished.

"Don't worry. I know just how far he'll stoop," Maya ribbed.

"I'll bet real far," Will replied kiddingly.

"Thanks pal," Jack retorted. "Once again, you've backed me up faithfully, like a true friend."

"Come on, buddy. We have work to do." They entered the *Phantom*.

"I'm not comfortable being away from the *Luna C*," Maya complained. "I don't see the point in taking the *Phantom*."

Will explained, "Because the *Luna C* is a Fleet ship and it would stand out. The *Phantom*, being an Attradean ship, might cause some confusion."

"Well, maybe you have a point. I still don't like it, though."

Will continued, "Shanna, you'll assist Maya on the flight deck. Jack and I will handle the cannons if needed."

"What happens when we get there?" Shanna questioned. When she noticed Will's hesitation, she added adamantly, "I'm not staying here on the ship while you go off and take care of business! I'm part of this team, too."

Will was prepared for some resistance from the girls over his plan to leave them on board. "Look, Shanna, you're pregnant. I don't want you getting hurt."

She became angry and warned, "If this baby is going to come between us, I don't want it!" She broke into tears and marched away.

"Do you have the coordinates for Gallia?" Will asked Maya, frustrated over Shanna's behavior.

"Yes I do. I looked into Fleet portals in that sector as well. I believe there are two that we can use."

"Excellent. Now I'd better go talk to Shanna."

"Good luck," said Jack. "She's a stubborn woman."

"Yeah, I know."

"Better you than me," he teased.

Maya grabbed Jack by the ear and pulled him toward the stairs. "Let's go before you put your foot in your mouth."

"Ouch! That hurts."

Maya looked back and winked at Will. Will chuckled to himself and went after Shanna.

Maya and Jack entered the flight deck and took their seats. "So, you and Will are going down to the surface alone," Maya remarked cynically.

"Yeah. That's the plan. If we get into trouble, I'm counting on you to get us out of there in a hurry."

"Don't worry. I'll be there for you," Maya assured him. She kissed him on the cheek. "Besides, I've got to watch over my property."

"Who says I'm your property?"

She nibbled on his ear and replied, "I do."

"Well, since you put it that way."

As Maya studied the controls, Jack pointed to the panel. "Over here is the control system for the transporter."

Maya joked, "Now how would you know about that?"

"Bastille taught me."

"You just tell me where it is and I'll operate it."

"Feeling a little cocky this morning, are we?" Jack teased.

"No more than usual," Maya retorted. Jack tilted the seat back and closed his eyes. Maya became annoyed and asked, "What do you think you're doing?"

"Taking a nap. You've got everything under control, right?"

"You are such a smart-ass."

• • • • • • • • • ● • • • • • • • • •

Will knocked lightly on Shanna's door and waited. There was no response. He turned the handle and opened the door. Shanna lay face down on the bed, sobbing. Will quietly closed the door and knelt next to the bed. He placed a hand on her shoulder. "Shanna, let's talk about this."

"About what?" she snapped. "You don't need me around anymore."

"That's not true. I would rather have you backing me up in battle."

Shanna rolled over and faced him. Her eyes were red and filled with tears. "Really, Will?"

"Of course."

"As much as I want to have the baby, I don't want to be stuck in the palace playing mother," she informed him. "It's not me."

"I know it's not you. All I'm saying is that it's safer for Jack and me to go down first. We have no idea what kind of danger lurks on Gallia."

"What if you get hurt?"

"Jack will do a good job. He and I are alike in a lot of ways, and I think it's important for him to do this as well."

"This time, I'll make an exception but, just so you know, it hurts me to be away from the action."

"Just think, once you've had the baby, it'll be like old times."

"But I can still fight. I'm not that far along."

"Yes, and you will fight. All I'm saying is that we don't know what we're up against. Just wait this out for me and see what happens."

"Alright. I'll try."

Will kissed her with a few short pecks on the lips. Shanna grabbed him by his shirt and said, "You can still kiss me like a man, Will. The baby won't mind." She pulled him close and kissed him enthusiastically. They rolled back onto the bed until Shanna was on top.

"Some things never change," Will kidded.

"Are you complaining?"

"No. Once again, it's just an observation."

Shanna placed her face close to his and said softly, "Shut up, you geek." They kissed again.

· · · · · · · · ●●●●●● ● ●●●●●● · · · · · ·

They ascended the stairs to the flight deck. Shanna wore the leather outfit that belonged to Will's stepmother, Tera. She strutted behind Will, adorned in the gothic battle attire that she became accustomed to. Will rapped on the door and entered. Shanna followed him with both of her hands on his waist.

Jack and Maya were focused on the monitor. Maya wore her snug, blue officer's uniform and black boots. Her blonde hair danced across her shoulders when she turned her head.

"Are we interrupting anything?" Will asked.

"Funny you should ask," Jack answered somberly.

"That doesn't sound good," Shanna responded.

"Looks like six unknowns are shadowing us," Maya told them.

"Are we cloaked?" Will asked.

"Now we are. I don't think they saw us."

"Any idea who they might be?"

"We're scanning them from long-range, but the distance makes it quite difficult," Maya commented.

"Care to take a guess?"

"They came from the thirteenth quadrant," Jack said.

"Do we know anyone there?" Will asked.

"How about GSS?" Maya suggested.

Will studied the monitor and remarked, "Well that's interesting."

"Perhaps GSS is the front for the Weevil," Jack suggested.

"Where would GSS have room for six ships of that size?" Shanna inquired.

Will answered, "Neva told me she thinks there could be a portal in the vicinity of GSS headquarters."

"You know, that would make sense," Maya responded. "The space station only has small ports for transportation purposes. Mynx and Neva confirmed that in an earlier discussion. A nearby portal would explain a lot of things."

"Do you think the station itself could be a front for a portal?" Will asked.

"Who knows? Maybe it is," Jack responded.

Maya added, "It has that large ring underneath that is supposed to be for orbit control. Maybe that really is a portal."

Will ordered, "Contact the *Leviathan* right away. I want to know where they are and if they've seen anything unusual." Maya entered a code into the computer.

"Is it serious?" Shanna asked.

Will remarked coldly, "I know what they're doing. This could be real ugly."

Maya called on the transmitter, "Come in, *Leviathan*. This is *Phantom*. Come in, *Leviathan*."

Will continued, "They suspect we're going to Gallia and are following us. They won't attack us until they see what happens to us."

"And?" Shanna asked anxiously.

"And they know we won't leave Yord unprotected. The *Luna C* is no match for more than two ships that size, so they know we'll have to bring the *Leviathan* out of hiding."

Maya repeated her call, "Come in, *Leviathan*. This is *Phantom*. Do you read us?"

"If the *Leviathan* fires, they'll pinpoint her position and swarm on her," Will surmised.

"So what can we do?" Jack asked.

Will watched the monitor and muttered, "Not a thing."

Celine's voice finally crackled from the monitor. "Go ahead, *Phantom*. It's Celine."

Maya pressed three buttons and Celine's face appeared on the monitor. Maya sighed with relief. "Boy, Celine, you had me worried for a moment."

"Sorry. We just got back and we're picking up a lot of unusual activity from an area in the thirteenth quadrant."

"Yes, from GSS."

"Ah, you recognized it, too."

"Can you see any ships coming or going?" Maya asked.

"Yes. Three Weevil cruisers," Celine replied.

"Can you tell where they go to or come from?"

"Oddly enough, it appears they are traveling through the ring underneath GSS," Celine said. Will paced the floor while the girls talked.

"I think Will has some instructions for you."

He knelt in front of the monitor. "Hi, Celine. How are things there?"

"Not bad. Someone wants to say hello."

Bastille's face appeared in the monitor. "Hello, Will."

Will was surprised and replied, "Bastille! How are you doing?"

"Not bad. My vision is returning."

"That's great. I'm glad to hear it."

"What's happening out there?"

"We have trailers behind us. Look, I think we're all being baited. Contact Emperor Rethos and ask him to send a small force to defend Yord in case of an attack. Tell him I've gone to Gallia and I'll explain when I get back."

"Sure. You don't want us to take care of business?"

"No. I think it's a trick to make us reveal the location of the *Leviathan*. Don't do anything unless it's absolutely necessary. If you do, the Weevil will likely hammer you with everything they have."

"The Weevil?"

"Yes, the Weevil. Just do what I tell you. It'll make sense later."

"No problem. We'll just observe."

Celine appeared on the monitor and interrupted him. "Anything else, boss?"

"Is Mynx or Neva on board?"

"They sure are."

"Ask them about the history of the space station. I don't think it's always been Galactic Security Services."

"Gotcha. Be safe."

"Thanks, Celine. We'll see you soon." The monitor went blank.

"You have a good idea what's going on, don't you?" Jack noted.

"I sense we're in for a big surprise. Things are starting to become clearer now."

"So what about our friends back there?" Maya asked.

"Let them follow us. Keep scanning their ships until we get a better idea what capabilities they have and who they are."

"How long before we arrive?" Shanna inquired.

"We've found two unknown portals that are active. If they match our coordinates, we'll be there in time for lunch," Maya said.

"Whose portals are they anyway?" Will asked her.

"No idea. They're there and they're open."

Jack added, "We've already passed through the Fleet portals. These are bonus."

"Maybe they want us to go to Gallia," Maya suggested.

"What if we're giving the Weevil too much credit?" Shanna asked.

"We covered that before with the Boromians when we thought they were behind everything," Will reminded her. Shanna squinted and focused on the floor.

Will immediately noticed. "You feel it, too, Shanna."

"Yes. The Eye is showing something evil lurking in darkness. We should go back."

"Are you sure, Shanna?" Maya asked.

"Yes, I'm sure."

"How is it that only you can sense the Eye in detail at this distance?" Will asked.

"Maybe her range is enhanced by her pregnancy," Maya suggested.

"Stop it. You're giving me the creeps," Will uttered

Shanna pouted, "I'm sorry, Will, but I'm frightened."

"We can't stop now. If we turn back, those ships will be all over us."

Maya informed them, "We've passed through the portal and the ships are still behind us."

"Are they in targeting range?" Will asked.

"Yes. Scanners show each ship is well-armed and easily capable of striking us at any time."

"Can you tell if the Weevil are on them?"

"I'm not sure. It doesn't appear so, but I could be wrong."

"Then who is it?" Will asked angrily.

Maya snapped at him, "I can't tell! Go sit down somewhere until I find out." Will apologized and left the cabin.

"I've never seen him like this," Shanna said.

"Stay here with Maya, Shanna," Jack ordered. "Will and I need to have a talk before we do this."

Jack left the cabin and pursued Will. He found him in the main quarters, sitting against the wall. "Are you alright?" he asked.

"I don't know."

"Are we going to do this?"

"Yes."

"Then let's get ready. If we're going to die, I want to die fighting like a man."

"You think we're going to die, Jack."

"I don't know. It doesn't sound very promising, though."

"I'm scared."

"Then what do we do about it?" Jack questioned him, concerned.

"We do what we came here for." Will stood and paced about the room.

"Didn't Mariel once tell you that you need to learn how to use your strengths?"

"Yeah, when Neva and I fought Nestor and the other Weevil. I got my butt whooped."

"Maybe we need to think tactfully here." Will waited impatiently for further explanation. "I don't think we can go in there with guns blazing," explained Jack. "Maybe you could shift into your alter-shape and scout the area first."

"What about you? You can shift, can't you?" Will asked, curious.

Jack stuttered, "I, uh, I don't know. I haven't tried to."

"Well, this could be your opportunity. At least we can identify the threat before we encounter it."

"Yeah, unless they eat life forms like us."

"Wow, I never thought of that," Will mentioned, growing more concerned.

"Please don't say anything to the girls about shifting. I don't want Maya to know if I do," Jack pleaded.

"Why?"

"I'm nervous, that's all. I've never seen her do it and she's very secretive about it."

"I understand. Your secret is safe with me."

Shanna descended the stairs and approached them. "What are you two babbling about?"

"Nothing important," Jack lied.

"Come on, Jack. I'm not that naïve."

"We were just discussing what might happen on Gallia," Will told her. "And?"

"We don't really know what to expect."

"Well, I thought you'd like to know we'll be there shortly and our friends have stopped following us."

"Is that good?" Jack asked.

"I don't know if it's good," responded Will, "but I expected it."

Shanna informed them, "We're going to leave the extraction coordinates active. If you need us, we'll be there. If you need to get out in a discrete hurry, be at the precise spot and you'll be transported back to the ship automatically."

"The ship can do that?" Jack asked.

"Maya says it has that capability." Shanna approached Jack and hugged him. Her eyes were wet. "Promise me you'll take care of Will."

Jack held her hands in his. "Of course. He's my best friend."

Will got a lump in his throat. "I didn't know we were best friends."

"We are. We just never get time to act like it."

"I'm sorry, Jack. If we get back..." Will paused. "When we get back, we're going to become a fighting unit: you, me, Shanna and Maya. We're going to kick the snot out of these Weevil assholes!"

Shanna complained, "I don't like this, Will, and if it was up to me, we'd go back. However, I support you and I'll do whatever you need me to."

"Thank you, Shanna. I appreciate that."

"So make sure you come back to me in one piece," she advised him. Will kissed her.

Jack quipped, "I think I'll, uh, check on Maya." Shanna was oblivious to Jack's remark and continued to kiss Will. Jack ascended the stairs and left them alone.

"I love you so much Will. I don't want our love to end in tragedy."

"It won't. We can overcome this."

"You'd better. We have a daughter to raise. There are lots of preparations to make."

"Like what?"

"Like making the universe safe for her."

"Ah, that's right."

"We can do this," she assured him. "It's our destiny."

"I'm glad you think so."

Maya's voice blared from the intercom, "We're in position for transport, Will. We're scanning the surface."

"I guess it's time," Will commented in a somber tone. Shanna pulled him close and hugged him.

Jack and Maya joined them. Maya's eyes were red from crying. Jack looked pale. Will gently pushed Shanna back and announced confidently, "Let's do this."

"That's more like it," replied Jack.

They stepped onto the transport platform. Will checked his sword and pistol. Jack checked both his pistols.

"Goodbye, girls," Will said sadly.

Jack waved and said humbly, "I love you, Maya. Take care, Shanna."

Maya pressed a knob on the local transport control panel. The men vanished in a flash. Shanna and Maya hugged each other and sobbed.

"They aren't coming back, are they, Maya?"

"Don't think like that. They are coming back." They ascended the stairs and returned to the flight deck.

· · · · · · · · ● · · · · · · · · · ·

The landscape on Gallia was boggy, with small trees and bushes scattered about. The air was warm and humid with light drizzle falling from the sky. Will and Jack appeared on the edge of a steep ravine. They instinctively drew their pistols and scanned the area.

"So far, so good," Jack remarked.

"I don't see anything either," Will confirmed

Jack pointed to a cave at the bottom of the ravine. "See that?"

"It's as good a start as any."

"I'll go down first. Cover me."

"Be careful."

Jack cautiously descended the side of the ravine. Halfway down, he paused and hunkered close to the ground. Will laid on his stomach and watched anxiously.

Two creatures galloped along the bottom of the ravine. They appeared to be horses until they approached the cave. Each had a hideous head with a pair of horns sprouting from it. Their skin was dark and leathery with frills of skin dangling from their underbellies. The faces were elongated like a horse's, but opened like the huge jaw of a crocodile with four tentacles wriggling about, each with a tiny eye at its end.

Jack was horrified at the sight of them. He laid close against the sloped ground to avoid detection. The creatures approached the cave and halted.

They sniffed the air and looked about suspiciously. He grew nervous and fidgeted. Suddenly, he lost his footing and slid down the side of the ravine.

"Jack!" Will screamed in horror.

The creatures roared and pawed at the ground as Jack slid toward them. He tried desperately to break his slide. Will drew his sword and went into a controlled slide down the side of the ravine.

Jack fired several shots at the creatures as he approached them. The first shot struck the leg of one of them. It yipped loudly and retreated a short distance. The other was persistent and snarled viciously at Jack. He stopped sliding about ten feet from the bottom and regained his balance. Quickly, he fired another shot.

The creature charged and plodded up the side of the ravine toward him. It bit into his boot and yanked on his leg. Jack lost his traction and slid downhill between the legs of the beast. It turned and snapped at Jack as he slid by, barely missing his arm.

Will slid down toward the unsuspecting creature and sliced its gut with his sword as he passed underneath it.

Further down was another creature, poised to lunge at Jack. He tumbled away from the creature and struggled to his feet. The creature knocked him down with its front hooves and was about to close its jaws on Jack's neck. Will slid toward it and fired a shot. He grazed the back of its head but didn't deter it.

Jack grabbed the lower jaw with one hand and tried to aim his pistol with the other. He shoved the pistol into the creature's gaping mouth but the creature clamped its jaws on the pistol and jerked it away from him. The pistol flew from its mouth and landed on the ground nearby. Jack grabbed its lower jaw and struggled to push away from it. The tentacled eyes whipped at him, leaving welts on his hands.

Will stowed his pistol and raised his sword. He jumped onto the back of the creature and slammed the hilt of his sword into its neck. The creature reared on its hind legs and roared.

Will grabbed three of the beast's tentacles and held on tight. The creature leaped about rabidly, trying to throw him off of its back. It raced into the cave with Will still riding on its back.

Jack retrieved his pistol and hurried after Will. Once inside the cave, he was surprised that he could see so well in the dark. For the first time, he was aware of the traits he acquired from Maya. When he reached an intersection in the cave, he saw a herd of the creatures racing toward him.

"Oh, shit!" he shouted and retreated behind a boulder.

Will rode the back of the creature into the depths of the cave. He ducked repeatedly to avoid rocks jutting from the low ceiling. They galloped under a steel catwalk toward a huge underground courtyard. Will reached up and grabbed onto the next catwalk as the creature raced past. He nearly fell as he pulled himself on top of the structure.

Three of the creatures nipped at his feet but missed. Will looked to the end of the catwalk and saw an iron door. At the other end was a ladder ascending the rocks to a steel deck, high above.

A low-pitched, wavering siren rang through the underground courtyard. The creatures immediately dispersed into several caves along the far wall. The iron door clanked and opened. Will instinctively rushed toward the ladder. He quickly undressed and transformed into his alter-shape, a wolverine. After the transformation, he nudged his clothes, sword and pistol behind a rock and then scurried for shelter.

Three big soldiers stormed through the door and spread out on the catwalk. Their faces and noses were partially covered with armor, as were most of their bodies. Large, yellow eyes with small dark pupils shone through their helmets.

One soldier searched the rocks and spotted Will in his wolverine shape. He laughed sadistically as he raised a short, silver bar and aimed at him. When he squeezed its handle, it fired a white pulse of energy in a blinding flash. Will fell between two rocks, out of their sight and landed with a yelp on his back. The soldiers laughed cynically and left through the iron door. The siren stopped and the courtyard became quiet.

Jack crept through the cave to the courtyard. He climbed up the rocks to the catwalk and scanned the area. He saw no sign of Will and approached the ladder. When he found Will's belongings, he whispered, "Will, are you here?" He leaned over the rocks and searched for him. Again he whispered, "Will. Are you down there?"

Will completed transforming back to his human form. He was naked and wedged between the rocks. "I'm here," he whispered. Give me a hand."

Jack climbed down the rocks. When he realized Will was naked, he turned away, red with embarrassment. He pulled Will up and avoided eye contact as Will crept past him. "Geez, Will! You're butt-naked."

"No kidding." Will gathered his belongings and quickly dressed.

Jack stared across the courtyard, still red-faced. Will teased, "What's the matter, you never saw a naked man before?"

"Well, I try not to make a habit of it. Oh, just get your clothes on already."

They climbed the ladder to the deck above. Jack scrutinized the courtyard and searched for any sign of the soldiers. Will focused on a dark tunnel at the rear of the deck.

Jack whispered, "All clear in the courtyard."

"Look at this – a tunnel," Will mentioned, pointing toward it.

As they approached it, Jack asked, "Do you sense anything?"

"Strangely enough, no. I sense absolutely nothing."

They entered the tunnel and followed it for a distance. Jack rubbed his hands along the wall. It felt strange to him. "Hey, Will, feel this."

Will ran a hand along the wall. He drew his sword and held it near the wall. The sword pulled toward it.

Jack removed his pistol and held it close. The wall attracted it as well. "It's magnetic."

Will stowed his sword and remarked, "I'll bet this is the ore we heard about."

"Why do you think they put a tunnel through it? It certainly doesn't look like they're mining it."

"What if this ore's a protective barrier around their base or dwellings?"

"Like how?"

"I don't know. Maybe that's how the Gallians became a mystery."

They came to a split in the tunnel. At the end of one, they saw a watertight door with a red light mounted on the wall above it. The end of the other tunnel was illuminated with a ghostly gray aura. They approached the lighted area.

· · · · · · · ●● ●● ● · · · · · · ·

Maya watched the monitor and completed another scan. Shanna paced about the flight deck nervously.

"You and Will are so much alike. He paces when he gets nervous, too," Maya commented.

"Something isn't right, Maya."

Maya's eyes widened. She recognized the images of the approaching ships as Fleet ships.

"You're not kidding something isn't right!"

"What is it, Maya?"

"Those ships are coming and they're Fleet ships! What the hell are they doing out here?"

"I have no idea."

"Did Steele know we were coming here?" asked Maya.

"I don't think so. We didn't get the vaults open until after she left."

They watched anxiously as the ships passed the cloaked *Phantom* and descended onto Gallia.

Maya blurted, "I don't believe this! Steele *is* up to something."

"Maybe we should go down there and warn the boys."

The mountainside opened wide and all six ships disappeared inside. Maya and Shanna stared at each other, stunned by the Fleet's arrival. Maya replied nervously, "Let's go find them. This is all wrong."

She set a timer on the transport control panel and then she and Shanna hurried down to the transporter platform. Maya warned, "Get ready for anything. This sucks already."

Shanna was about to speak but a bright flash interrupted her. She blinked and found herself on the surface of Gallia. Maya stood behind her and tapped her on the shoulder. Shanna jumped. "Oh, Maya! You scared me."

"I see an opening on top of that ridge." She pointed halfway up the mountainside. They hiked to the top and were haunted by an eerie silence.

"I think they know we're here, Maya."

"How can you tell - a vision?"

"No. This place is supposed to be so formidable, yet we haven't seen any sign of resistance."

When they reached the ridge, they discovered a small, steel platform manned by two Gallian sentries. The girls hid in the bushes and observed them.

"They look pretty mean," Shanna whispered.

"I don't think they're human. Look at the size of them."

"I'll handle this." Shanna climbed out of the bushes and pulled herself onto the platform. Maya froze in horror as she fretted over what might happen.

The sentries were surprised to see Shanna and aimed their weapons - two short silver rods strapped to their knuckles.

Shanna pointed to the opening in the mountain and stared at them. She attempted to communicate telepathically but to no avail. She held up a finger for patience and reached down to her knife straps.

The sentries were curious until she drew two of her knives. She held them by the blade and offered the pair to them. They looked at each other baffled. Neither could understand why a petite woman like Shanna was there in the first place; let alone what harm she might do them. They each reached for a knife.

Shanna quickly raised her hands and fired the knives into their eye sockets. They fell to the ground, screaming in pain. Shanna quickly drew another knife and cut each one's throat. She retrieved her knives, wiped them clean on the sentries' boots, and then stowed them in the straps on her thighs.

Maya climbed onto the platform, visibly shaken. "Shanna, what the hell was that all about?"

"Nothing really. Let's go."

Maya wasn't used to following orders from someone smaller or younger and was stunned by Shanna's bold behavior. They opened a thick, metal door behind the platform and entered a long tunnel. There was a faint glow at the end.

Maya grabbed Shanna's shoulder and whispered, "Let me lead in case we encounter Fleet officers. You cover me." Shanna stepped aside and allowed Maya to pass.

At the end of the tunnel, they stood by the railing of an underground hangar. At the bottom were legions of Gallian soldiers in perfect formation. They appeared to be waiting for someone important. At the rear of the hangar were the six Fleet ships. The hatches opened and a party of eight emerged from each of the ships.

Maya muttered in a low tone, "What the hell is going on here?"

"They look like real Fleet Officers," whispered Shanna.

Maya studied them and noticed one in particular at the head of the party. She ordered Shanna, "You keep an eye for Will and Jack. I can feel them nearby."

"I can, too. I'll find them."

Shanna sensed anger in the tone of Maya's voice. She left her and crept along the rail. Finally, she spotted the men across the hangar from them. She returned to Maya and informed her of their location.

"Don't try to contact Will yet. We don't want to risk giving away our presence." They observed, anxious to know more. The leader of the Fleet officers was dressed in black and wore a dark mask.

"If I didn't know better, I'd swear that was a woman," Shanna commented.

Maya replied tersely, "And a woman we both know."

Shanna's eyes grew wide with surprise. "You don't think that's Steele, do you?"

"As a matter of fact, I know that's Steele."

"What do you think she's up to?"

"I don't know, but she'd better have a good explanation."

· · · · · · · · ● · · · · · · · · · · ·

Will and Jack crept closer to the hangar. They kept their attention focused on the Fleet members, anxious to find out why they came to Gallia.

Steele stood before a formation of soldiers and studied them. A soldier barked an order to his men. The soldiers stomped their feet twice and froze in attention.

Steele addressed them in a strong, aggressive voice: "We are nearing the day when we'll dominate the Orion quadrants. You will no longer live in recluse under the surface of your world. You will expand and colonize other worlds with your race, crushing anyone who stands in your way."

The soldiers again stomped their feet twice and chanted, "Hail, Cherenka."

Steele held her hands out for silence and continued, "We have weeded out the Weevil from Fleet headquarters and regained control of the region. The Boromians and Attradeans are but a shell of their former selves. As soon as the *Leviathan* is in our grasp, we will launch our assault on the

Weevil from both sides of the Omega Portal." She walked to a vacant corner and met with several Gallian officers.

One of them asked, "Are you sure the *Leviathan* has been located?"

"Oh, yes, and it's there for the taking."

"But how did you find it, Commander Cherenka?"

"Believe it or not, I was invited to see it."

"You will be richly rewarded if we succeed."

"Oh, don't worry, Major, we will succeed. I will return for the *Leviathan* as soon as we finish our business."

"Then let's not delay our success." They passed through a doorway and disappeared.

Will looked at Jack in disbelief. Jack chastised Will, "I told you she was a bitch, but *no*, we had to give her a chance."

Will replied reluctantly, "Fine. You were right about her. Now, let's get out of here." They scurried toward the front of the hangar.

· · · · · · · ● · · · · · · · · · ·

Maya's face was beet-red with anger. "Are you okay?" Shanna asked.

"I'm going to kill her. She betrayed us all. She used me to get to the *Leviathan*."

"Come on, Maya. We've got to go back and warn the others."

A soldier pointed at them and shouted orders in Gallian. Five soldiers quickly surrounded Maya. Shanna darted away before they could catch her.

One soldier punched Maya in the face and knocked her to the ground. She got to her knees and spit blood on his boot. He kicked her in the side of her head. Maya fell to the ground, dazed and helpless. Two soldiers picked her up by her arms and carried her away.

Shanna climbed up the rocks and hid. She searched for Maya but couldn't see where they took her. Suddenly, she was grabbed from behind and a hand covered her mouth. She struggled desperately, but was pulled back onto the platform. When she was turned around, she was face-to-face with Will. "Be quiet," he instructed her.

Will and Jack released their hold on Shanna. She glared at them. "You guys nearly scared me to death."

"Sorry, but we had to make sure you didn't scream."

"Did you see where they took Maya?" Jack asked.

Shanna pointed to the rear of the hangar. "They went that way, but I lost sight of them."

"There's going to be some ass-kicking if they hurt her," Jack declared.

Shanna concentrated for a moment. "What is it, Shanna?" Will asked.

"I know where she is." Shanna rushed across the platform to a ladder with the men right behind her. She descended the ladder with the speed of a squirrel, racing down a tree, leaving the men lagging behind. They came upon another steel door and paused. She placed her hands against the door and concentrated again.

"Will I be able to do that?" Jack asked.

"I can't even do that." Will quipped.

"It's clear," Shanna said. "Come on."

Will and Jack pushed on the door while Shanna pulled on a thick chain. The latch on the other side grated as it moved. The door opened and they entered a long, dark hall.

"Can you sense anything, Shanna?" Will asked.

"Yes, I can. We're getting close."

· · · · · · ● · · · · · · · · ·

Steele and a Gallian officer, Major Duowd, sat at a fancy glass table made of prisms. Beams of colored light struck the table from various openings in the walls. The reflections created a psychedelic effect.

The officer poured dark, red liquid into a glass and served it to Steele. He then filled a glass for himself as well. They held their glasses toward each other for a toast. The Major announced, "A toast to you, Commander Cherenka. You have delivered us to the brink of salvation."

Steele replied humbly, "A small price for what we both stand to gain."

They tapped their glasses and sipped greedily. The major noticed Steele's look of satisfaction and remarked, "I see you approve."

"Yes, I do. They say that the blood of your victims is the sweetest nectar."

"I agree, but we believe that you also consume the soul of your enemies when you drink their blood, thus gaining their strength."

Two soldiers stopped outside the door, holding Maya by her arms. One rapped on the door. The studs on his gloves made a metallic pinging sound.

Major Duowd stood and politely excused himself. He approached the door and opened it. "What is the meaning of this interruption?" he demanded.

One soldier explained, "We caught this woman in the hangar. I believe she was spying on us."

Steele got up and approached the door. "Let me see your spy."

The soldiers stood Maya up and pushed her through the doorway. Major Duowd grabbed Maya's arm and bent it behind her back.

Surprised, Steele asked, "Maya, what are you doing here?"

"Funny, Steele. I was about to ask you the same thing."

"Why does she call you Steele?" Major Duowd asked.

"Oh, just an old nickname."

Steele grabbed Maya by the chin and turned her face upward. She grinned sadistically as she examined Maya's wounds. "It's ironic, Maya, how you always lived by the book. Look what it got you."

"Why are you doing this, Steele?" Maya muttered.

"Because I can. Look at those pathetic excuses for officers in the Fleet. They don't deserve to live."

"So who made you judge and jury?"

"I did," Steele replied arrogantly.

Major Duowd asked, "What would you like us to do with her?"

"Feed her to the Equinostriae."

"This isn't over, Steele," Maya uttered defiantly.

Steele wiped blood from Maya's cheek and licked it from her finger. "You taste good enough to eat. I'm sure the Equines will be more than pleased with you." She gestured with a wave of her hand for the soldiers to take her away.

Full of rage, Maya shouted, "I'll stop you, Steele! I swear!"

Steele punched her in the face and staggered her. When Major Duowd closed the door, she glared at him with a long cold stare. "How did she get here?"

"I... I don't know."

"I'm sure she didn't come here alone. Have your men search the base from top to bottom."

"Yes, Commander Cherenka."

Will, Jack and Shanna peeked around a corner. Two soldiers dragged Maya toward them. Jack nudged Shanna behind him. His face tensed as he prepared to attack. The soldiers turned the corner and were surprised by Jack. He rammed into the chest of the first one and knocked him over.

He ripped off the armored helmet and pounded mercilessly on the alien soldier's leathery face. Will drew his sword and slit the other's throat, just under the armored mouth guard.

Maya groaned as she hit the stone floor. Shanna lifted her up and held on to her. "We've got to stop Steele," Maya said weakly.

Shanna urged her, "Save your strength. We're getting out of here."

Will pulled Jack off the soldier. "That's enough, Jack. We've got to go."

Jack drew his pistol and shot the soldier in the face. The face was obliterated, exposing remnants of the oversized skull. Jack stowed his pistol and declared "Now it's enough." He took Maya from Shanna and carried her. They followed the tunnel back into the hangar.

· · · · · · · · ● · · · · · · · · · · ·

Major Duowd opened the door for Steele. She left the room and immediately noticed the dead guards at the end of the corridor. She bellowed, "Son of a…!"

"What is it?" asked Major Duowd.

"She got away! Find her!"

The Major urgently activated a pull box on the wall. Sounds of multiple sirens, both high and low-pitched, filled the air.

Will and Shanna rushed ahead of Jack and Maya. They crossed the hangar to the nearest Fleet ship. Shanna entered and initiated the ship's controls for take-off. Will held off several of the soldiers with shots from his pulse pistol. Jack saw the number of soldiers approaching and turned back.

"Come on Jack" Will shouted.

"We'll never make it!" Jack replied. He carried Maya behind a large stone, which jutted out from the bay wall.

Will swore loudly, "Damn it, Jack! Come on!" Several soldiers fired at him.

Will ducked inside the ship and closed the hatch. He bit his lip while pausing at the hatch. "Damn it, Jack," he repeated. "Why didn't you keep going?" He rushed up the stairs to the flight deck.

Shanna asked, anxious, "Did they make it alright?"

Will sat down and initiated the power sequence on the ship. He replied somberly, "No, they didn't." The ship shuddered as a growing number of soldiers fired at them.

"What happened, Will?"

Three soldiers approached the front of the ship with large shoulder-mounted grenade launchers. "Not now, Shanna!" Will increased power and spun the ship around. The tail of the ship sliced the soldiers in half. The others scattered. He steered toward the exit of the bay but the steel doors were closed.

Shanna fretted, "Now what?" Will turned toward the small cave.

"What are you doing, Will? We'll never fit through there!"

"We don't have to. Fasten your seatbelt."

"Will, you're insane!" Shanna cried. Will accelerated to the tunnel and spun the ship sideways. The ship slammed into the tunnel, jolting Will out of his seat. His arm caught on a panel door and tore open. Blood streamed from his wound and pooled on the floor. Shanna bounced hard against the side of the cabin and was stunned from the impact.

Will quickly got up and pulled Shanna to her feet. "Come on, Shanna!" He ordered. "We've got to move." He took her hand and led her down the stairs to the hatch. He pressed a button and waited impatiently for the door to open. When it finally did, he pulled Shanna out behind him.

Shanna stumbled and fell. She cried, "I can't run, Will. Something's wrong." Will picked her up and carried her through the tunnel. Shanna sobbed, "Something's really wrong, Will."

"No kidding," he muttered angrily. They exited the cave and ascended the rocky slope.

Shanna repeated, "I mean it, Will - the baby."

Will breathed heavily as he struggled to reach the top of the ravine. He asked urgently, "Where do we go for the transporter?"

Shanna pointed to an area between two large rocks. Will staggered as he covered the last steps to the rocks. He set Shanna down and fell to his knees. "Now what?"

Tears streamed down Shanna's cheeks. "Give it a few seconds to activate."

Will looked at Shanna and realized she was in trouble. "What is it, Shanna?"

"The baby. I think she's hurt."

"What do you mean?"

Shanna sobbed emotionally and hollered, "What don't you understand, Will?" A flash interrupted Shanna's plea. They arrived inside the *Phantom* on the transport platform. Shanna fell to the floor, unconscious.

Will cried out, "Shanna!" He picked her up and took her into the nearest cabin. When he set her down on the bed, he saw she was bleeding badly.

"Don't give up, Shanna. I'll get help." He kissed her forehead and left the cabin. Will raced up to the flight deck and activated the flight systems. As soon as the starting sequence finished, he flew the *Phantom* away from Gallia.

· · · · · · · ●●● ●●● · · · · · · ·

Jack held Maya tightly as he peeked from behind the rocks. Maya whispered, "What happened to Will and Shanna?"

"I don't know?"

Maya struggled to sit up but Jack blocked her. "Stay put until I tell you."

"I'll be fine, Jack. What happened to them?"

"I told you already. I don't know."

Maya glared at him. "They are my family, Jack."

"Their ship crashed against the wall," Jack revealed. "I couldn't tell if they got out."

"What are we going to do?"

"Can you walk?"

"Yeah, I can walk."

"The soldiers are tying lines to the ship. It looks like they're going to try and move it away from the tunnel."

"Any sign of Steele?"

"No. Why?"

"I want her dead."

"Let's just concentrate on getting out of here." Jack searched the area at the rear of the hangar and revealed, "There's a platform over there with a doorway. Let's give it a shot."

Maya fought back tears and pulled Jack to her. She hugged him tightly. "I love you, Jack."

"I know," he said, forcing a smile. He knew that they were likely going to die and there was nothing they could do about it.

Jack helped Maya to her feet. He took her by the hand and led her across the floor. They climbed the short steel ladder to the platform and disappeared in the darkness.

IX

TURNING THE TABLES

Bastille and Celine sat at the controls on the flight deck of the *Leviathan*. Bastille explained, "I still think the Weevil are being helped by someone inside the Fleet."

"Well, that shouldn't be an issue anymore, since Steele cleaned house at Fleet Headquarters," she explained. The monitor beeped three times and Bastille pressed the 'receive' button.

Will's frantic voice pierced the calm. "Celine! Bastille! Are you there?" They noted the urgency in Will's voice. Celine answered, "We're here, Will. What's wrong?"

"Steele is the leader of the Gallian fighting force. Maya and Jack are trapped on Gallia and Shanna's hurt badly."

"What do you want us to do?" Bastille asked.

"Take the *Leviathan* to Gallia. See if you can pick up Jack or Maya on the scanners. There's a magnetic substrata so you may have difficulty."

"We're on the way."

Will added, "Don't make a move until I get back. I have to get help for Shanna."

"Good luck, Will," Celine said. "She's in our prayers."

The *Leviathan* sped toward Gallia at maximum speed.

Will placed the *Phantom* on auto-pilot. He descended the steps to the lower level and entered Shanna's cabin. She was burning up with fever so he soaked a tee-shirt in the sink and placed it across her forehead. "Come on, Shanna. Hang in there."

Shanna grabbed his hand with both of hers and pleaded, "Please, Will, do something."

"We'll be on Yord in a little bit. You've got to hold on."

Will placed his hands on Shanna's stomach. He could feel movement within her womb. Suddenly, he had an idea. "Shanna, can you channel your healing powers through me."

"What for?"

"Maybe you can heal yourself and the baby through me."

Shanna struggled to concentrate. She placed her hands on Will's head and groaned in agony. Will focused on the baby and kept both hands flat on Shanna's womb. He felt a tingling in his palms. "Something's happening, Shanna!"

Shanna screamed and strained with every ounce of energy she could muster for close to an hour. She heaved and gasped with every breath. Will suddenly worried that she might do more harm than good. "Ease back, Shanna. I think it's working."

Shanna took a deep breath and lay back on the bed as sweat poured over her face. She exhaled spoke wearily, "The baby's going to be alright."

Relieved, Will bowed his head and then kissed Shanna's cheek. "Get some rest. I'm taking you home."

· · · · · ● ● ● ● ● ● ● ● ● ● ● · · · ·

The *Phantom* docked in the underground facility on Yord. The hatch opened and Will carried Shanna out through the hatch. Arasthmus and Mariel waited anxiously but when they saw Shanna their joy turned to fear. Mariel asked fretfully, "What happened, Will?"

"Shanna was injured."

"Follow me," Arasthmus ordered. Will carried Shanna up the stairs, behind Mariel and Arasthmus.

Mariel guided them into an ornate bedroom at the end of the hall. Will laid Shanna gently on the bed. Mariel immediately noticed the dry blood stains on Shanna's legs and fretted, "Oh, no! She was bleeding."

Will explained, "She channeled her healing powers through me to save the baby." Two servants entered the room and stood by the door.

Mariel ordered, "Find Keira and bring her here." One of the servants promptly left the room.

Will warned, "Steele is preparing the Gallians for war. They plan to take over this and several other quadrants."

"I thought Steele was on our side," Arasthmus remarked.

"Steele is on Steele's side. The only two obstacles for her are the *Leviathan* and the Weevil."

"That complicates things quite a bit."

"Worse yet, she knows the base and she knows we have *Leviathan*."

"Rethus' ships are keeping an eye on Yord," Arasthmus informed him.

"Yeah, but they don't know that the Fleet ships are bad guys."

"I can take care of that."

"Arasthmus?"

"Yes."

"I need you to return to Gallia with me. I have to rescue Jack and Maya."

"Just tell me when. An old guy like me needs a little action every once in a while." When Mariel stared angrily at him, Arasthmus replied innocently, "I didn't mean anything by that, Mariel."

"I'm sure you didn't."

Keira and Regent entered the room. Mariel exclaimed, "Thank goodness you're here! Shanna's had an accident and the baby could be in jeopardy."

Keira replied, "I'll do what I can."

Mariel informed her, "I'm going with the men to rescue Maya and Jack. Take care of her."

Arasthmus was surprised. "What? You're going, too?"

"I can fight and I can shift as well. Besides, Steele's going to pay for putting Shanna in danger." Mariel stormed out of the room.

Arasthmus looked at Will nervously. "I never saw her like that before," Will said.

"Me neither.

"Can you join us?" Will asked Regent.

"Of course."

Keira looked at Regent with concern. He hugged her and assured her they would be okay. Keira kissed him and turned her attention to Shanna.

The men hurried to the transport bay. When they reached the *Phantom*, Mariel was already on board. She barked impatiently, "Let's go, already!"

Surprised by her take charge attitude, they boarded quickly and departed.

· · · · · · · ●· · · · · · · · · ·

Jack remarked, "Those traits of yours sure come in handy."

Maya kidded, "I wondered how long it would take you to use them." Jack opened a door at the rear of the platform and peeked in. "What do you see?"

"A long tunnel, sloping downward."

"Should we try it?"

Their conversation was interrupted by Steele's loud voice booming across the hangar. "Find them or else I'll execute you all, one at a time."

Major Duowd and four soldiers escorted Steele from the wreckage at the tunnel entrance. He informed her, "They can't escape. There's no place to go."

"If the drifter and the officer are here, so are the other two ragtag criminals. I want them all and I want them now!"

Major Duowd bellowed several orders in Gallian. Groups of soldiers dispersed in different directions. "We can handle this little problem, Commander. I don't want to hold up your schedule."

Steele turned and poked a finger in his face. "I can't afford any loose ends. These people are cunning."

"I understand."

"I will check back with you by day's end and you'd better have solved this little problem. If not, you can forget the remaining components for the Tridents."

"I understand, Commander." Steele then spoke into a small communicator on her wrist.

Jack and Maya watched from the rocks above. Jack whispered, "Can you hear what she's saying?"

"No, but they're returning to the ships. It looks like they're leaving."

Major Duowd bellowed more orders and pointed to their platform. Jack muttered, "Oh, shit. They see us."

"We'd better get moving." They passed through the doorway and hurried down the hall.

"Do you think they'll come back for us?" Jack asked.

"I know they'll come back for us. We just have to stay alive until they do."

The end of the tunnel grew brighter. "I think there's a light up ahead, Jack."

"Yeah, maybe it's a train."

"What do you mean by that?"

"Oh, nothing."

They crept close to the side of the tunnel and watched anxiously for the source of the light. Three flashes filled the tunnel. One of them struck Jack in the arm. He groaned and fell to the ground, clutching his arm in pain.

Maya helped him up and urged, "Come on, Jack. We've got to keep moving." She could hear the rapid footsteps of the approaching soldiers.

Three more flashes lit up the tunnel. One struck Maya in the leg and another grazed her head. She moaned briefly and fell to the ground unconscious. Jack lay next to her. He placed his arms around her and hugged her tightly. "Maya! Say something, please."

Blood seeped from the open wound on the side of her head. A patch of her long blonde hair had burnt away along with part of her scalp. The soldiers surrounded them. One punched Jack in the head with a studded, leather glove. He laid unconscious, bleeding profusely from his wound.

· · · · · · · · ● · · · · · · · · · ·

As Will piloted the *Phantom* back toward Gallia, Arasthmus sat in the co-pilot's seat, examining the controls. "Did you learn to fly one of these at the academy?" he asked.

Will cracked a smile and replied, "No. I learned from watching everyone else."

"Not bad."

Mariel and Regent entered the flight deck. "Do you have any idea yet how we're going to execute this rescue?" Mariel asked. Will thought, but said nothing.

"Do you think they're still alive?" Arasthmus asked.

"I can only hope so. I can't leave them if there's any chance they're still alive."

"What do you want us to do?" Mariel asked.

"I need Arasthmus on the ship in case either of them needs medical attention."

Arasthmus replied, "I'll be ready whether it's shipboard or on the planet's surface."

"Regent, I could use you to watch my back," continued Will.

Mariel waited impatiently for her task. Will continued monitoring the long-range scanners. "What about me, Will?" Mariel finally asked.

"What do you want to do?"

"I'm going down there with you."

"Mariel, I don't think that's a good idea."

"Will, don't tell me what's a good idea. It was your decision to come here in the first place."

"I know, but..."

"But what?" Mariel asked.

"Alright! Answer one question for me."

"What is it?" she asked defensively,

"What do you know about GSS?"

"GSS? Isn't that your Galactic Security Services?"

"Yes, but did GSS stand for anything else that you know of prior?"

"Prior to what?"

"Prior to Tenemon's arrival. What about when Yord was an economic metropolis?"

Mariel thought for a moment. "Come to think of it there was a wealthy group of shippers from Gallia, operating over several quadrants. I believe they used the name Gallian Shipping Services but I never associated that with GSS."

"What ever happened to them?"

"They just disappeared."

"I wonder if the Weevil or Tenemon had anything to do with it."

Arasthmus commented, "I seem to remember hearing an earlier conversation where you said the Weevil controlled the GSS space station."

"Yeah, I did."

"You thought there was a portal underneath or behind the station."

"Yes. I did."

"What if that was the method the real GSS used for their shipping industry? If the Weevil took it from them, it would pretty much ruin them."

"But why would they disappear?"

"Perhaps when Tenemon came on the scene, they withdrew into seclusion."

Will considered the scenario and then his eyes widened. "Until one day when they are strong enough to retake their space station and reestablish their livelihood."

Regent commented enthusiastically, "I think we're getting somewhere."

Mariel asked, "So how does that help us now?"

Will explained, "They follow Steele out of necessity. She's helping them prepare for their independence. They hide under the surface of the planet and build a military force. No one can see them nor confirm their existence because of the magnetic barrier around them."

Arasthmus concluded, "That would explain why the Weevil are so anxious to stop them."

"Yes, and it would also explain why the Weevil on board the Boromian ship wanted to know where Steele was," Will surmised.

"But why is the *Leviathan* such a big deal?" Arasthmus asked him.

"If the space station generates portals to any part of the universe, it can be used to send armies as well as to conduct commerce. The *Leviathan* would protect the station and make it virtually untouchable."

A beeping sound caught Will's attention. When he checked the monitor, it showed a green wall with five blips approaching.

"What is it?" Mariel asked.

"We're passing through the portal. There are some ships coming toward us." Will reached across to the armament panel and set it up to fire five torpedoes.

Arasthmus asked, "Do you think it's wise to get into a dogfight out here?"

"Perhaps you're right, Arasthmus. Those are the Fleet's ships. My guess is that Steele's is the lead ship."

"So?"

"So, the *Leviathan* is waiting on the other side. Once Steele enters the portal, the *Leviathan* can destroy the other four. She'll never know what happened to them."

"What if she comes back?"

"She still won't know what happened. Her scanners didn't record any cannon fire, so by all accounts, she'll have no idea who did what."

"I think you're dreaming, Will," Mariel chided. "That's really reaching for a miracle."

"We'll be clear of the portal soon. I'll show you." He activated the ship's cloaking system and cut the ship's speed to fifty-percent. He pressed several buttons and sent a beacon. After a lengthy pause, he received a return signal. He activated the monitor to transmit. "Come in, Celine. This is *Phantom*."

Celine's voice came back quickly. "Go ahead *Phantom*."

"See those five ships?"

"Yes. They just passed us a short while ago."

"Are they still in weapons range?"

"Yes they are."

"When the lead ship enters the portal, take out the last four."

"After the lead ship enters the portal, you want me to take out the last four?" she repeated.

"That's correct, Celine. Timing is everything."

"Will do."

"We're out." Two clicks echoed back. "Now we wait," Will announced optimistically.

Everyone's eyes were glued to the monitor as the lead ship approached the portal. "Steady. Steady," Will murmured.

The lead ship entered the green wall on the monitor and disappeared. Four torpedoes streaked from the *Leviathan* but struck only two of the other four ships. They erupted in brief balls of flame. The remaining ships veered away from the portal.

"Damn!" Will shouted. He pressed three switches on the armament panel and fired five torpedoes. He watched nervously as the ships tried evasive maneuvers but, as three of the five torpedoes found their mark, the remaining ships exploded.

Will jumped up and screamed, "Gotcha!" He immediately pushed the *Phantom* to full power and raced toward Gallia.

"You've got to be the luckiest person alive," Mariel chided.

Arasthmus kidded, "Or the dumbest son of a bitch."

"I'll take lucky," Will said. "It was a calculated move and I was prepared for the possibility that some of the ships might dodge the torpedoes."

Arasthmus conceded, "I'll give you credit for that. Maybe it was luck and possibly a little skill."

Will eyed the monitor. "We're closing in on Gallia."

"How come no one else has been able to land on Gallia and survive?" Mariel asked.

"Because we're not landing. We're transporting down to the surface," explained Will.

"Just that simple," she remarked sarcastically.

"Well, not exactly. There are some creatures out there that might pose a problem."

"What do you mean might?"

"Well, the Gallians locked them up when we were inside the hangar."

"I'm sorry I asked," she groaned.

Will cut power and programmed the transport system. "Arasthmus, I need you to monitor us carefully in case I need you."

"I'll be ready."

Will stood and walked to the door. He paused and announced to Mariel and Regent, "It's show time." They proceeded to the transport bay.

Will stepped onto the transport platform and looked back at them. "Well, are you coming or not?" Mariel and Regent looked at each other hesitantly for a second and then stepped onto the platform. Will counted down, "Three, two, one."

A bright flash illuminated the room. Instantly, Will, Mariel and Regent appeared on the surface of Gallia, near the cave opening.

"So far, so good," Regent remarked.

Will stepped carefully down the embankment toward the cave's entrance. Regent followed behind him. They paused near the bottom and looked for any sign of the Gallian creatures but they saw nothing. Will looked up the slope but there was no sign of Mariel.

"Mariel! What the hell?" Will shouted and rushed back up the slope. He searched the area but saw no sign of her.

"See anything?" Regent hollered.

Will shook his head and descended the slope again. "She pulled one over on me, Regent."

"What do you mean?"

"She's got her own plan."

"Mariel is pretty shrewd."

"Remind me to never underestimate her again."

They quietly entered the cave. Regent hung onto Will's belt as they passed through the darkness. The wreckage from the Fleet ship was pushed off to one side away from the hangar's entrance. They entered the hangar and slipped behind it for cover. Six soldiers passed by and climbed the ladder to the catwalk.

"Come on," Will whispered, "We've got to move fast."

"What for?"

"Just come on!" They raced toward a second ladder near the rocks, climbed halfway up and stopped.

"What's wrong?" Regent asked.

"The soldiers. They haven't left yet."

"So?"

"They'll see us if we go higher."

One soldier mumbled into a small transmitter. The hangar echoed with the grating sound of metal. All of the cage doors around the perimeter of the bay opened, releasing the equinostrians. The soldiers finally left the catwalk and closed the door behind them.

Will ordered Regent, "Hurry!"

As they continued their climb, one of the beasts spotted them and lunged at the ladder. Regent grabbed a rung tightly and yanked his legs up just high enough to avoid the creature's jaws. He hurriedly climbed the remainder of the rungs.

"Are you okay?" asked Will.

Regent breathed a sigh of relief. "Yeah. Boy, that was scary."

"Stay close," Will advised him.

"You don't worry about that."

Will hurried across the platform and passed through the iron doorway. Regent drew his pistol and followed closely behind.

· · · · · · ●●●● ● ●●●● · · · · · ·

Inside a dank room, Jack was strapped in a wooden chair by leather belts. Metal wires were taped around his arms and legs. The wires were tied to a knife switch mounted on the side of a transformer.

Maya was strapped to a short table. Her legs hung down and her feet rested in a bucket of water. Next to her bed was another knife switch mounted to a transformer. Its wires wrapped around her arms, legs and head. Major Duowd entered with three of his soldiers. He ordered one of the men, "Wake him up."

The soldier shook Jack several times, but Jack was delirious and incoherent. The soldier picked up a bucket of water and dumped it over Jack's head. He groaned as he opened his eyes. Major Duowd pulled up a chair and sat across from him.

"What do you want from us?" Jack asked.

The Major laughed at him and mentioned, "I was about to ask you the same thing."

"Let us go, please."

"Let's see. You trespass on my planet and destroy a ship on my premises. Why should I let you go?"

Jack struggled to hold his head up. "Please," he begged.

Major Duowd nodded to one of the soldiers, who punched Jack in the side of the face. Blood flew from his mouth and spattered onto the wall as he whimpered in pain.

The Major asked, "What does your lady officer want here?"

"She doesn't want anything from you."

"Shall we ask her?"

"Leave her alone," Jack pleaded. The Major nodded again to the soldier. The soldier shook her several times but she didn't respond.

Major Duowd commented cynically, "I think she needs a little wake up call." The soldier closed the knife switch and Maya was electrified. She screamed and convulsed. Her body became erect as an arrow.

Jack shouted, "Please, leave her alone!"

The soldier opened the switch and Maya's body lay limp. Small plumes of smoke rose from the burns on her skin along with the odor of burning flesh. The Major inquired calmly, "Now, what is she doing here?"

"She worked for Steele Furey. We came here looking for answers and discovered that Steele was working with you guys."

"Who is Steele Furey?"

"Your Commander."

"That is Commander Cherenka. What does your lady officer want with Cherenka?" Jack looked down at the ground in silence.

The Major nodded. The soldier closed the switch near Jack. Jack shook wildly. The chair bounced a foot across the floor. He screamed in agony as the wires burned into his skin. The soldier opened the switch. Jack sobbed and his skin smoldered as the shock subsided.

Will and Regent stood outside one of the doors in the hallway. Will drew his sword and stood near the door handle. He whispered to Regent, "They're in here." Before Regent could answer, Will knocked on the door.

Regent was shocked and whispered, "What are you doing, Will?"

"Just watch."

The Major ordered one of the soldiers to check the door. When the man opened it, Will pulled him out and stuck his sword under the soldier's helmet and through his throat. The soldier fell to the ground clutching desperately at his throat.

Will stormed into the room and warned, "I'd advise you boys to get away from those transformers." The Major held his hand up for the soldiers to stay put.

Will threatened, "We can do this the easy way or the hard way."

One soldier crept toward Will's left. Regent stepped through the doorway and surprised him. He leaped on the soldier's back and held his pulse pistol in the man's face. The soldier fell to his knees from Regent's weight on his back. Another soldier reached for the switch on the transformer near Maya.

Will drew his pulse pistol and fired a shot at the transformer. He quickly hurdled the bed and struck at the soldier's arm with his sword. The bone in the soldier's arm made a crunching sound as the sword broke through the bone. The soldier fell to the ground, holding the bloody stump. Will became annoyed with the man's screaming and fired a shot into his face, killing him instantly. Will asked sarcastically, "Who's next?" The last soldier backed away from Jack's transformer.

Will ordered him, "Release both prisoners."

"Don't you dare," Major Duowd countermanded.

Will pointed his pistol at the soldier, who quickly began to unbuckle the straps on Jack's arms and legs. After he finished, Will instructed him, "Unstrap the woman as well."

Regent pushed his captive further into the room. "Sit down and don't move or else," Will warned them.

"You'll never get away with this!" The Major yelled angrily.

Will replied, "I think you need a lesson in manners. Get in the chair."

"No!" shouted Major Duowd. "I will not!"

Will fired a shot at the Major's foot. He fell to the ground, writhing in pain. Regent pulled him to his feet and shoved him into the chair. He secured the Gallian with the leather belts. Will asked Jack, "Are you okay?"

"I…, I think so. But Maya…" Jack went to her and placed his head on her chest. He cried for her.

"Keep a gun on those two, Regent." Will pointed to the soldiers and then went to the transformer. He placed his finger delicately across the switch. The Major closed his eyes and cringed, but nothing happened. "That's right, Major. I could kill you right now but I won't."

"What do you want from me?"

"First, a little Q&A. Did your people run the old Gallian Shipping Services?"

"Yes, but that was a long time ago."

"What happened to it?"

"Why do you care?"

"I care because I'm returning things to the way they used to be," Will informed him.

"What are you talking about?"

"Yord was one of Gallia's biggest customers, weren't they?"

"Yord is gone. It's just a memory, just like Gallia Shipping Services."

"Did Gallia use the GSS space station to control their shipping routes?"

"Yes, but that was before those bastard Weevil took it from us. We solicited the Attradeans for help and they destroyed our empire. King Tenemon took our colonies one by one until this was all we had left."

"How many of you are there?"

"None of your business. Why should you care that we live beneath the surface of our last remaining planet like animals in squalor?"

"We can help you regain your space station. Yord is reborn and Gallia can be also."

"I don't believe you."

"Let us walk out of here and I will personally recover your space station."

"That's it. Just let you walk out of here and you won't kill me?"

"It's my token of good faith, even after the treatment you gave my friends."

"Fine. Let me go and I'll escort you out of here." Will unstrapped him and pointed to the door.

The soldier asked Will, "Can you really make things the way they used to be?"

"If not, I'll die trying."

The Major walked cautiously ahead of them. Will advised the soldiers, "Don't make a ruckus or people will die in the confusion." They nodded in agreement.

Regent carried Maya from the table to the corridor. Jack staggered behind them.

The Major asked, "What do you have to do with Commander Cherenka?"

"We befriended her. We helped her recover her command from the Weevil and she betrayed us," Will explained.

"What of this *Leviathan* she seeks?"

"There is no *Leviathan*. It's a myth we generated to protect us in the early stages of Yord's reformation."

When they reached the hangar, the Major ordered the creatures locked down. A siren sounded and the creatures disappeared into their pens. The gates closed and the siren ceased.

"I'll go down the ladder first, then you," Will instructed him. The Major stared passively as Will descended the ladder and backed away from the base. "Come on, Major. Your turn."

The Major grimaced and climbed down the ladder, as blood dripped from his injured foot. Several soldiers gathered nearby. Will ordered the Major to put them at ease.

Major Duowd bellowed, "Everything's under control. Let them be." The soldiers were surprised, but heeded their leader's order.

Will marched him through the tunnel to the slope outside. "Well, Major, you can return. Thank you for your cooperation."

Regent spoke into his transmitter, "Arasthmus, we're coming up."

"I'll stay here until Mariel shows up," Will informed them.

"We're not going back without you."

"Jack and Maya need help. That's an order."

Regent reluctantly obeyed. He spoke into his transmitter, "Arasthmus, bring three of us up." They were quickly whisked up to the *Phantom*.

The Major smiled at Will and remarked, "I think you grossly underestimated me, Will Saris."

"I see you know my name."

Twenty soldiers emerged from the tunnel and watched anxiously. The Major drew a pistol from underneath his armor chest plate. "Cherenka said you would be here. You are so predictable."

"Kill me and you kill any chance of restoring Gallia's heritage."

"Do you think I care about Gallia's heritage?"

The soldiers raised their guns and aimed at Will. Will became annoyed with the outcome of his ploy. He inquired, "What do we gain if we kill each other?"

The Major laughed and mocked him, "You'll be shredded before you get that shot off. I assure you."

A frightening snarl sounded from the rocks overhead and then a gryphon pounced on the Major, placing its talons at his throat.

Will received a telepathic message from the creature; "It's me, Mariel. I'll cover you."

Will beamed at the Major and said, "Let's see, you can order your men to drop their weapons and retreat or I can order the gryphon to rip you to pieces."

The Major cringed from the gryphon's weight on his chest. "You mean to tell me this is your creature?"

"It's one of many." Will looked at the gryphon and quipped, "Mariel, they need some convincing."

The gryphon pressed its talon harder against the major's throat. Blood seeped from a small cut and ran down the side of his neck as he trembled. "Alright! I surrender."

"Then order their retreat."

The Major commanded the soldiers to put down their weapons and retreat into the cave. They balked and maintained their positions. "Now!" Duowd snapped. "Do I make myself clear?" One by one the soldiers reluctantly put down their weapons and retreated into the cave."

Will picked up the Major's pistol and pocketed it. "It's a shame we couldn't work things out." The gryphon climbed off the Major and flew up the slope into the trees.

"The next time we meet, I'll kill you. I swear," warned Major Duowd.

"The next time we meet, you'll be in shackles."

"Screw you, Saris!"

Mariel descended the slope and joined Will. "Tenemon used to say that and look where it got him," Will remarked disappointedly.

His comment surprised the Major. "What do you mean? What happened to Tenemon?"

"His ill humor caused him to not be among the living anymore."

The Major was amused and commented, "Oh, don't worry, Saris. I'll make sure you join him."

Mariel pulled on Will's arm and said, "That's enough. Let's go."

"Once we take care of Steele or Cherenka or whatever she calls herself, you won't have a leg to stand on, Major."

Mariel bellowed, "Will, that's enough!"

"Come on, Saris. You and me in a duel."

Mariel took Will's arm and activated his transmitter. "Arasthmus bring us up, now," she ordered.

Will continued to goad the Major. "I dueled with Tenemon. Why wouldn't I duel with you?"

Mariel covered Will's mouth. She warned the Major, "You are at a dangerous crossroad, Major. Pick your friends carefully."

She and Will were transported back on board the *Phantom*. Mariel removed her hand from Will's mouth and glared at him. Will replied sheepishly, "Sorry, Mariel. I got carried away."

"You buffoon! The man is nearly twice your size and three times as strong and you want to fight him in a duel? He'd rip you apart!"

"How do you know he's three times stronger?" Will asked, embarrassed.

"Because I had to hold him down. It wasn't as easy as it looked."

"I'm sorry."

"I told you about using your skills and not your big mouth."

"Alright! I'm sorry."

"This is why I came. I let you go your own way just to prove my point. You can't rule as a king when you behave like this."

Will had nothing left to say and looked dejectedly at the ground. Mariel put an arm around him and advised, "Go check on your friends."

X

RECOVERY

Will walked somberly down the hall from the main quarters. He overheard Arasthmus talking with Jack. "There'll be no way of knowing how bad the damage is for at least three days. She's been through a lot."

Will sat on the floor outside the room and cried. He felt responsible for what happened to both Jack and Maya. Jack came out of the room with his head down. Will looked up at him with teary eyes and uttered, "I'm so sorry, Jack."

"Me, too. We really screwed up big time on this one."

Will wiped his eyes. "No, I screwed up on this one. How is Maya?"

"She's still unconscious. Her vitals are pretty good, according to Arasthmus, but she might have other problems."

"What kind of other problems?"

"We don't know if the electric shocks did any damage to her organs or her mind."

"I've got to see her," Will muttered and edged past Jack into the room. Jack sat on the floor and he, too, sobbed.

Arasthmus sat next to Maya and dressed the wound on her head. He glanced at Will and then resumed his task. Will sat on the other side of Maya and took her hand in his. He tried to communicate with her telepathically, but there was no response. He rubbed her hand against his cheek and prayed for her to recover.

"Don't you think you should take us back?" Arasthmus asked. Will looked drained, and unconcerned by the question. "I know how you feel but this doesn't change the fact of the matter. We need to get back to Yord," he continued.

Will replied humbly, "I'm so sorry, Arasthmus. Let me know if there's any change in her."

Will left him and stopped in the hall by Jack. "When you feel better, come up to the flight deck. We need to talk."

"Give me about five minutes," answered Jack. "You have an idea?"

"I do. Things are going to be different from now on."

"I wondered how long it would take before this war hardened your spirit," Jack commented.

"It's not only hardened me. It's given me a mean streak that I never had before."

Will proceeded to the flight deck. When he was situated in the pilot's seat, he stared at the lime-colored orb that was Gallia. Will uttered angrily, "You guys want a war? You got it." He powered up the ship and cruised toward Gallia.

A few minutes later, Jack entered and sat down in the co-pilot's seat. He took one look at the monitor and was stunned. "What are you doing, Will?"

"Leaving a calling card."

The *Phantom* entered the atmosphere and approached the mountains. Will explained, "That mountainside is the entrance to their hangar." Jack was mesmerized and said nothing.

Will activated the ship's armament control panel and programmed three torpedoes. He locked onto the mountainside and prepared to fire. A beeping sound distracted him. "What's that?" he asked.

Jack checked the short-range sensors. The monitor changed and he saw a large, radar-shaped device rising out of the landscape. "I think we're in trouble, Will."

"Screw it!" Will shouted. He launched the torpedoes and veered away. In the distance, he saw a second radar-shaped device rising out of the landscape. "Now it's time to cut and run."

He pressed the *Phantom* to full power and raced to the outer atmosphere. "Jack, set the long-range scanners to aft and find out what the hell those things are."

"I'm on it." Jack entered several codes into the computer and set the monitor for a second channel. The monitor displayed Gallia's landscape. A bright red and yellow burst of flames erupted from the mountainside. "Nice shooting, Will."

"I couldn't miss. Bastille improved the tracking system on the torpedoes. Once they lock on, they can't deviate." He reset the panel and programmed three more torpedoes.

Jack asked curiously, "What are you doing now?"

Will spotted three of the radar-shaped devices on the monitor. He moved the arrow on the monitor to each one and depressed a switch each time. "I'm leaving them a present," he answered, feeling arrogant.

The ship lurched and slowed down significantly. Jack grabbed the armrests tightly. "What's happening, Will?"

Will fired the three torpedoes and waited patiently. "Those dishes create a magnetic field that draws ships to the surface. My guess is that they're strong enough to slam a ship to the ground. After that, the creatures do the rest."

A moment later, all three dishes exploded and the *Phantom*'s speed increased a little more.

"Shouldn't we be going faster?" Jack asked.

"There're more of those dishes locked onto us."

"Then fire more torpedoes."

"Can't. That was the last of them."

"Oh, great!"

"It's only slowing us down. We've got a good head start on them."

The monitor beeped and three LEDs lit up on the control panel. Jack switched the monitor's channel. A lone Fleet ship was approaching them.

Will checked the instrumentation for the cloaking system. The display read "100% cloaking in effect" and a green light indicated no malfunctions.

Jack quipped, "Looks like Cinderella's back again."

"Perfect. They'll have to shut down the fields." At that moment, the *Phantom* accelerated.

"Right on cue. Huh, Jack?"

"You are so damn lucky, Will."

"Nah, just pissed off."

The monitor beeped again and flashed two red lights. Will checked the display. "Looks like the *Leviathan* is calling." He keyed the receiver. "Go ahead *Leviathan*."

Celine's face appeared on the monitor. "Did you rescue them, Will?"

"Yes, we did. Maya's in bad shape, though."

"What do you need us to do?"

"Go on back. I think we stifled Steele's plans for a little while."

"See you at the dock." Will clicked the 'send' button twice in response. The monitor reflected a wavering image as they approached a green wall of light.

"There's our portal," Will told Jack.

"You mean 'There's Steele's portal.'"

"Yeah, for now."

Mariel entered the cabin and stared at Will. Jack sensed her anger and commented, "I'm going to check on Maya." He quickly left the flight deck.

Will looked at her innocently and asked, "What?"

"You know what? You always have to have the last word, don't you?"

"What are you talking about?"

"The torpedoes. Why did you have to do that?"

"What would you have done?"

"Maya is your priority right now. You get lost in too many details."

Will frowned and stared straight ahead. Mariel knelt next to him and placed a hand on the back of his neck. "Look, Will, I'm trying to help you."

"You have a funny way of showing it," he replied sarcastically.

"You're taking too many risks."

"Yeah, like risking my life to stay on Gallia to look for you."

"I was fine. I was watching over you," she replied.

"I didn't know that. I would give my life to save you and everyone else that's close to me."

"I'm sorry. You did stand your ground when the others left."

"You know, Mariel, that was pretty interesting," Will remarked and cracked a smile.

"What was?"

"You – a gryphon."

"I didn't pick it."

"I know. I think it's a magnificent creature."

"Thank you."

"How's Maya doing?" he asked.

"No change."

Will fretted, "I'm really worried about her."

"Yes, she's in bad shape."

"What about Shanna? Do you think she'll be alright?"

"I don't know. Look, when Tenemon invaded and hurt my friends and family, I became a warrior. When the fighting was over, I became Mariel the overseer of the priestesses again."

"I didn't know you could fight."

"I choose not to, unless I'm provoked."

"The future of Yord is in danger. I felt this was necessary," explained Will.

"And what have you learned from this trip?"

"The Weevil are waiting to invade the quadrant from another part of the universe," Will told her.

"You're sure?"

"Yes, I am. Secondly, the Gallians want revenge for their suffering since the Weevil took their portal station from them and Steele is willing to help them."

"Why does that bother you?"

"Because she's selling out the Fleet for personal gain. She wants to get her hands on the *Leviathan* as well. If she does that, she'd be unstoppable."

"Why do you fear that Yord is a target?"

"Because that's where the *Leviathan* is kept. She'll kill everyone until she gets what she wants."

"Do you have any idea how to stop all this from happening?"

"I think we need to find out what's going on at the GSS space station first. That could make a difference in which way we have to move."

"I will fight by your side."

"Mariel, you don't have to do that."

"Yes, I do. You and Shanna are my family. If you fight, I will fight."

"We'll see if it comes to that."

The ship shuddered and a bright light whisked past the *Phantom*.

"What the hell was that?" Will yelled. The monitor beeped twice and Will acknowledged the call.

Celine's face appeared on the monitor. "Hi, Will. That was just the *Leviathan* passing you by."

"Well you scared the daylights out of me."

"Bastille and I wanted to see how fast she'd go, after the new modifications."

"And?"

"She's very impressive. See you back at the dock." Will watched in awe as the ball of light vanished ahead of them.

Mariel kidded, "Maybe it's time to trade this baby in."

"Yeah, you can say that again."

· · · · · · · · ● · · · · · · · · · · ·

With both ships docked inside the base on Yord, Will left the *Phantom* and proceeded up the stairs.

Mariel called to him, "Wait, Will." He paused. "Did you try to communicate with her?"

"No. I'm too nervous." They hurried up to Shanna's room. As Will entered, he feared what he might see.

Shanna sat up in bed and cried out, "Will!" Her hair was matted and flat. Her eyes were red and sunken and her face looked ashen.

Will looked at her and realized the agony she must have been through. "Shanna! Are you alright?"

She looked into his eyes and cried. "I was so worried about you. I could sense your emotions and I knew you were in trouble."

Will hugged her and assured her, "It's okay, now."

"What happened to Maya?"

"She's hurt badly."

"I can't feel her."

"She's not responding so far to anything Arasthmus has tried."

Mariel hugged Shanna tightly and cried, "Oh, my little girl."

"Thank goodness for Keira. She was wonderful," Shanna praised.

"What happened?" Mariel asked.

Keira entered the room holding a tiny infant girl wrapped in a blanket. "It looks like Will is a father," she announced.

Will was stunned. He gazed at the infant proudly. "She's so tiny!" Keira gently placed the infant in his arms. "But... but you weren't due for a while, Shanna," Will stammered.

"She's just like her mother and father - impatient and stubborn."

"What do we call her?" asked Will.

"How about Marina?"

"Marina it is."

Keira asked, "Is Regent okay?"

"Yes, he's fine."

"Thank goodness."

Will gazed at Marina and noted, "She's got your eyes, Shanna."

"I hope she's not as ambitious as Shanna," Mariel teased.

"Keep the boys away from her, just in case," Will warned.

"What are you two insinuating?" Shanna asked defensively.

"Nothing," Will teased.

"If I remember correctly, Mariel, the apple doesn't fall far from the tree," Shanna joked.

Mariel blushed. "That's enough, Shanna."

Will handed the baby to Shanna. Keira explained, "She's a bit early, so we need to take especially good care of her."

"When can I get out of bed?" Shanna asked.

"Not for at least a few days."

"What about Will?"

"What about him?"

"Can he get in?"

Keira's face reddened and she replied softly. "I've got to see Regent. I'll be back." She anxiously left the room.

Mariel chastised Shanna, "That wasn't nice. You know Keira is very reserved."

"I didn't mean to embarrass her. I just asked the question."

Will interrupted, "Shanna, a lot has happened. Jack is pretty messed up and Maya's in a coma. She was hurt really bad."

"I'm sorry, Will. I was so excited about the baby and seeing you again."

"I know. Please be patient for a bit."

"I really missed you, Will." Will leaned over and kissed her. Shanna grabbed his collar and pulled him to her. She kissed him ardently.

Will pushed away reluctantly and warned, "Easy, girl. Our daughter is watching." Shanna giggled and tickled Marina. She gazed as the infant smiled.

Will brushed the hair away from her eyes and said, "I'm going to check on Maya. Arasthmus may need help with her."

"Let me know if I can do anything. Remember, I am a healer now."

"Thanks, Shanna."

· · · · · · · · ● · · · · · · · · · ·

Arasthmus stepped out of the room and reached down to Jack. "Why don't we have a seat and talk?" Jack took his hand and stood up.

"Thanks for trying to help, Arasthmus."

"I wish I could make a difference in her." They walked into the main quarters and sat at the table.

"What are her chances?" asked Jack.

"Not good."

Will entered the main quarters through the hatch. Jack looked up at him sadly. "How's Shanna?" Arasthmus asked.

"She had the baby early."

"Is she alright?"

"Yeah. She looks a little worn out."

Will walked into Maya's room and knelt next to her. He took her hand in his and looked intently at her. He thought about their childhood and how close they were. He recalled the night she picked him up from the dance at the academy. She chastised him for being attracted to her. He recalled her words, "Will, we're like family."

Will heard her words echo through his brain, over and over. He stood up angrily and shouted, "There's got to be something we can do."

Jack burst into the room, followed by Arasthmus. "What happened?"

"There must be something Shanna and Mariel can do for Maya." Sensing Will needed time, Arasthmus departed quietly.

"If you have any ideas, I'm listening. I want her back as badly as you do," Jack reminded him. Will paced the room and desperately searched for an answer. "So what do we do now?" he asked Will. "We know about Steele's plot. We know about the Gallians. Where do we go from here?"

Will frowned and stared at Maya. "I can't just look at her like this. It's killing me."

"Me, too."

Will sat down again and placed his head in his hands. He sighed and spoke in a muffled tone. "We've got to find out what's going on inside the GSS station. This time, I really need a plan."

"Perhaps Mynx and Neva can help."

"Let's go talk to them."

Will and Jack stood up and approached the doorway. As they stepped into the hallway, they met Talia, Neva and Mynx. Talia looked past them and saw Maya. "How is she?"

Will and Jack stared vacantly at her. Talia pushed past them and sat next to Maya. "Maya, it's me," she pleaded. "Say something." Talia noticed the section of scalp and hair missing from the side of Maya's head. She asked emotionally, "Who did this to her?"

Jack answered, "The Gallians."

"They'll pay for this!" she sobbed.

Will approached Mynx and Neva. "Got time for a few words?"

Mynx replied, "Sure. Where at?" Will nodded toward the main quarters. They followed him to the table and sat down.

Mynx mentioned somberly, "We're sorry about Maya."

"It doesn't look good for her."

"What happened to her?" Neva asked.

"It seems the Gallians like to torture their prisoners with water and electricity."

"Good. When we get our hands around a few of their necks, I'll remember that's how they like it."

"You'll get your chance. I promise."

"What's on your mind?" Mynx asked.

"The GSS space station."

"What about it?"

"I hear it used to belong to the Gallians under the name Galactic Shipping Services."

"I've never heard of it."

"Me, neither," Neva added.

"The Weevil captured it from the Gallians and have controlled it ever since."

Neva and Mynx grew wide-eyed. "Are you sure about that?" Mynx asked.

"Absolutely. The Gallians used portals controlled by the space station to control their shipments. When the Weevil got hold of it, the Gallians lost their sole industry."

"So all the time we worked for GSS, we worked for the Weevil?" she asked.

"That's right."

Neva swore, "I'll get every one of those bastards. We were among the enemy all that time."

"Have you been able to crack the transmission codes?"

Neva answered, "Bastille thinks he's close. It's a very complex pattern they're using."

"If we can tap into their computers, we might be able to see what's going on." Mynx suggested.

"Can you do it?" Will asked.

Neva explained, "It would be difficult and lengthy to develop a link from one of our ships to the station. The most efficient way would be to plant a routing device on board, configured for us to access their database."

"That would be suicide," Mynx argued. "The station appears to be well protected, based on our infra-red imaging, taken while you were on Gallia."

"Where's Bastille now?" Will asked.

"He's playing with another invention of his on board the *Leviathan*."

"We need a very good plan for this mission. The danger has grown past anything we're ready for. Worse yet, I don't want anyone else getting hurt."

"How is Shanna?" Mynx asked.

"Funny you should ask. She had a baby girl." The girls congratulated him.

Mynx asked skeptically, "Wasn't that a bit quick for her to have the baby?"

Will chuckled. "You see, Firenghian women bear children in half the time human women do. The baby did come quite prematurely even by Firenghian standards."

"Is she okay?"

"Keira seems to think so. She's so small, though."

Neva asked, "What did you name her?"

"Marina."

"Ah, what a wicked name. I'd be glad to teach her a few things when she's older."

"She's my daughter, remember - not a galactic boxing champ," Will kidded.

Neva looked hurt. "I just thought you'd like her to know how to defend herself."

"We'll see."

"Why don't we take a walk over and see Bastille?" Mynx suggested.

"Good idea."

"I'll wait here, if you don't mind," Jack commented.

"That's fine. Let me know if anything changes," Will requested. They left the *Phantom* and crossed the dock to the *Leviathan*.

"Is Jack alright?" Mynx asked. "He looks like hell."

"The Gallians tortured him, too. He's pretty strong, but he's devastated by Maya's condition."

Will entered the *Leviathan* first. He gazed at the intricate control panels and hologram maps positioned on the first floor. "Wow, I didn't realize how advanced this ship really is."

"You haven't been on it before?" Neva asked.

"Only when we were stealing it and just long enough to secure it."

"Oh, there is so much to see." Mynx led them down the stairs to a room filled with electronic and electrical panels. "Bastille! Where are you?"

A voice echoed from the far end of the room, "Down here."

"Boy, it's really chilly in here," Will remarked.

"It's got to be to keep these complex systems operating," Neva explained. They walked down the main aisle until they came to an open man-way.

Neva shouted, "Are you down there, Bastille?"

Bastille's head popped up from the opening. He wore a baseball cap turned backwards and a flannel shirt with the sleeves rolled up. "Will! How are you?"

"I guess I'm alright, Bastille. How are you doing?"

"Much better now. I can play with my toys again."

"So what are you up to?"

"Tweaking the main drives. I think we can pick up the max speed about three percent."

"I heard you're making headway with the Weevil codes."

"Yeah, I think I'll have them solved in the next couple days."

"Great. There's something else I want to discuss with you."

Bastille pulled himself out of the man way. "Sure, let's talk."

"You made the torpedoes cloak."

"Yeah. It wasn't hard."

"Can you do that to a person?"

Mynx and Neva's mouths fell open. "Are you serious?" Mynx asked.

Will looked straight-faced at Bastille. "Can you?"

Bastille scratched his head. "Well, I never really thought about it. I'm not sure."

"It's very important. Think about it."

"What's more important, the codes or going cloaked."

"Both. I believe Neva and Mynx can help you with the codes."

"We'll figure it out," Mynx responded. "Now beat it before I change my mind."

Will winked at her and said, "I knew I could count on you." He placed his hand on Bastille's shoulder. "Thanks, Bastille. You don't know how important this is to me."

"Oh, I've heard. You have a knack of getting into some serious situations."

"You can say that again. Let me know if you have any new developments."

When Will departed the ship, he found Jack waiting patiently on the dock for him. "Everything alright?"

"Same as before."

"What brings you out here?"

"Will, we've got to do something. I can't sit around like this."

"But, Jack, you've been through a lot, yourself. You ought to try and get some rest."

"I can't. I'm so wound up over Maya that I feel like I'll explode."

"I'm upset about her, too."

"There's got to be something we can do." They walked back to the *Phantom*.

Mariel met them at the hatch and stated confidently, "Will, I have an idea."

"About what, Mariel?"

"About Maya. You mentioned that Shanna channeled some of her healing power through you."

"Yes, and it seemed to work pretty well."

"Perhaps Shanna, Keira and I could channel our powers together and attempt to heal Maya."

"Do you think Shanna's up to it? I mean she just gave birth."

"We can try. She's a strong, young woman."

"What's the worst thing that could happen?" Jack asked.

Mariel became serious. "If we do nothing, Maya could become undead. To leave her like that could lead to brain deterioration. On the other hand, the shock of trying to jar her mind into activity could kill her as well."

"Have you ever done this before?"

"After Tenemon destroyed our world and left, we tried to help some of the wounded. Those with head traumas this severe didn't fare too well."

"So this is a long shot?"

"Either choice is a long shot."

Jack looked to Will for hope. "They have very strong healing powers alone. I can't imagine how effective they could be collectively," Will mentioned optimistically.

"Shanna has uncanny power compared to most of our Seers and Healers," Mariel commented. "She may be the difference compared to our past attempts."

Jack paced the hatchway nervously. "I can't let her fade away," he moaned.

"Give them a chance, Jack," Will suggested.

"It's not my choice. You have a say. Talia has a say."

"No, Jack. Her heart belongs to you. It's your choice."

Jack placed his hands on his hips and looked up at the ceiling. "Damn! This isn't fair." He wiped tears from his eyes. "Please, Mariel. Don't lose her."

"I'll do my best, as will the others." She left the men alone.

"I know how you feel, Jack. We've got to believe she'll pull through." Will's voice was soft, comforting.

"That's easy for you to say."

Will grew angry with Jack's remark. "You think so? I love her, too! If her heart didn't already belong to you, she might have been mine. So don't tell me it's easy for me to say."

Jack looked at Will with an amazed stare. "What did you say?"

"You heard me. When I learned that it was you, I was happy for her. As much as I wanted it to be me, I was really happy for her. When I got to know you, I was glad it was you that won her heart."

Jack leaned against the wall and gazed at the ceiling again. "I'm sorry, Will. I didn't know."

"There's a lot you don't know. I'm your friend Jack. Do you think I'd be so selfish?"

"No, you wouldn't. I'm sorry."

Will walked to the black trunk in the corner. He took two gray crates off of it and set them down nearby while Jack watched. Will dragged the trunk toward the table and sat down next to it.

"What's that, Will?"

"It's the trunk with my dad's belongings." He unlatched it and opened the lid.

"Are you looking for something in particular?"

"I don't know. Maybe peace of mind." He removed a black box with silver trim and set it on the table.

Jack opened the box and removed a brown sack. Will cautioned him, "Be careful with that. I don't know what it's for."

Jack pulled the top of the sack back, revealing a golden skull. "What the hell is this?" Jack exclaimed.

Will glanced at it and remarked, "That's odd."

He retrieved a small electronic device with wrist straps, multicolored buttons and two small monitors. He set it on the table. A tag hung from one of the straps. It read, "Consult Xerxes before operating."

Will set several articles of clothing on the table, reached down for a small pile of papers and noticed an envelope addressed to him. His face became pale as he held it in front of him.

Jack noticed Will's uneasy look and asked, "What is it?"

Will stared coldly at the envelope. He stuttered as he replied, "It's... It's a letter. It's... from my dad." He continued to stare at it.

"Aren't you going to open it?"

"I don't know." Will's hands trembled.

"Maybe it's important. Maybe he knew something was going to happen to him."

"Do you think so?"

"Hell, yeah! He didn't just leave you a casual 'sayonara'."

Will opened the envelope and read the letter. "Well?" Jack asked impatiently.

"Yeah, he knew they were being set up. A friend of his helped arrange for me to avoid their fate."

Jack waited anxiously for more. "And…"

Will reached for the electronic device and strapped it to his wrist.

Jack asked nervously, "What are you doing, Will? What is that thing?"

Will read the letter again. He pressed three buttons on the device and pointed it at one of the walls in the main chamber.

"I hope you know what you're doing, Will."

A beam of light emerged from the device and painted a blue circle on the wall. A second later, the circle blurred.

Jack was stunned by the image. "Holy cow! What is it?"

"Our answer, I hope," Will replied enthusiastically. He paused in front of the circle and asked, "Are you coming?"

Jack was confused and reluctant since, once again, he had no idea what Will was doing. Will walked into the blue circle and disappeared.

Jack complained, "Oh, I hate when he does this." He cautiously followed Will.

· · · · · · ● ● ● ● ● ● ● · · · · · · ·

Mariel returned to the main quarters of the *Phantom* with Arasthmus, Bastille and Celine. When she saw the blue circle on the wall, she immediately studied the contents of the trunk on the table and read the note. She shuddered as she crumpled the note, tossed it in the trunk and frowned. Arasthmus grew concerned and asked, "What is it, Mariel?"

"Nothing, I hope."

"That didn't look like nothing." Bastille said.

"I just hope Will knows what he's doing."

Celine remarked, "I don't think there's anything he can't figure out."

"That's exactly why things are getting worse," Mariel said with anger in her voice. "He's going deeper each time into things we aren't ready to handle."

"I didn't think of it that way."

"We are in no position to take on any more problems."

Arasthmus studied the circle and then mentioned, "That's a portal. I've seen one of them before."

Bastille reached into it and watched his hand disappear. "I believe you're right."

"No one goes through there," Mariel warned.

"What are you afraid of?" Celine asked.

Mariel stared angrily at her again. "Can you guard that portal?"

"Of course. Why?"

"Arasthmus and Bastille, you'll carry Maya up to Shanna's room. I'll meet you there." They obediently went to Maya's cabin.

Mariel explained, "There's too much going on right now. We have to get past Maya first before we do anything else. If that portal gives any of our enemies access to this base, we're defenseless. That means Maya dies. That means the *Leviathan* falls into the wrong hands. That means we all are dead. Do you understand?"

Celine cowered, but replied, "I do. I'll guard the portal."

"If you see anything suspicious, you contact us right away and let me know. I'll send Mynx and Neva down here to help you."

Arasthmus and Bastille carried Maya's unconscious body on a stretcher from the *Phantom*.

Mariel calmed down and explained, "Look, Will is still immature in a lot of ways. He is brave and adventurous, but he doesn't realize how big this is. We have to put the reigns on him before we all get killed."

Celine replied humbly, "I didn't realize there was so much at stake, Mariel."

"If you only knew the half of it." Mariel hurried from the *Phantom* and followed the men.

Celine pushed the clothes and skull from the table into the trunk. After turning over the table, she took a defensive position behind it with her pistol drawn, ready to fire.

· · · · · · · · ● · · · · · · · · · ·

Will emerged from the blue portal and stepped into a conference room. No one was present as he peered into the hall. Two men with clipboards and white coats walked toward him. He ducked behind the door and listened.

"Xerxes hopes to find a place for us to escape to before we are overrun," one man said.

"Isn't there any way to stop the Gallians and the Weevil?"

"How? When Commander Furey defected, all of our weaknesses were exposed."

"I hope someone kills that bitch, slowly and painfully."

"You know, Scotty, I'd do it myself if I could, even if it meant dying."

Will watched from behind the door as they passed. Jack came through the portal and tapped him on the shoulder. Will jumped and whispered hoarsely, "What is wrong with you?" Jack grinned at him. "Follow me and don't make a sound," Will ordered. They followed the men down the hall until they entered an elevator.

Will watched, frustrated, as the doors closed. "What do you think?" Jack didn't reply. Will turned around to see an older, albino man holding a gun to Jack's head with a hand over his mouth. Jack could only shrug his shoulders.

Will asked, "Who are you?"

"You first and you'd better have a good answer. If not, you're both dead."

Will slowly set his pistol on the ground. "We didn't come here to harm you." The man stared at him somberly.

Will heard a voice inside his head that said, "Talk to me fast, boy." He realized the man was communicating telepathically with him. He responded, "I'm Will Saris and my friend is Jack Fleming."

Jack followed the communication and interjected, "I can think for myself, Will."

The man was surprised and spoke openly, "Will Saris, huh?" He released his hold on Jack and pushed him toward Will. He continued to aim his pistol at them.

"I sure am. Do I know you?"

"Maybe. Who was your mother?"

"Seneca."

"What was your mother?"

Will became defensive. "I hope you mean that in a respectful way."

"Just answer the question."

"She was Queen of Firenghia."

"And your father?"

"William Brock. He was from Earth."

The man extended his hand in friendship and said, "It took you long enough to get here. Where's the portal?"

Will shook his hand and asked, "How do I know you really knew my mother?"

"That's easy. She was a shape-shifter. She was killed, battling the Boromians."

"So who exactly are you?"

"I'm Xerxes. I've been expecting you. Now, where is the portal?"

"Down the hall in the room on the right."

"Take me to it."

Will and Jack led Xerxes down the hall to the conference room. "I came here to seek answers," Will told Xerxes. "I found a note in my father's trunk that instructed me to use this device to seek you."

"My boy, we have much to talk about and time isn't exactly on our side right now." They entered the conference room and stood near the blue portal.

"How much do you know about the war?" Xerxes asked.

"Where do you want me to start?"

Xerxes sat down at the table and directed Will and Jack to follow suit. Will answered, "I heard one of your workers mention Imperial General Furey. I know quite a bit about her."

"Talk to me."

"She's in charge of the Gallians and the Fleet. They are planning something big."

Xerxes muttered impatiently, "Yes, yes. We know that."

"They need to get the GSS space station back from the Weevil before they can proceed."

"Ah, GSS. I should have known she was after that. Do you know anything about a huge ship called *Leviathan*?" Will and Jack chuckled.

Xerxes was slightly embarrassed and inquired indignantly, "What's so funny?"

"Why is everyone worried about the *Leviathan*?"

Xerxes answered, "You don't know the importance of that ship, do you?"

"Of course I do. It's supposed to be the most powerful ship in the universe, a one of a kind man-o-war." Now it was Xerxes' turn to chuckle.

Will looked puzzled and asked humbly, "Alright, Xerxes. What am I missing?"

"You don't have all the answers, do you?"

Will was embarrassed and replied, "Fine. I'm listening."

"A long time ago, there was a group called the Council of Guardians. They controlled several power sources that, when in close proximity to each other, became the most powerful objects known at that time."

Will and Jack wondered where Xerxes was going with his revelation. Jack tapped his foot rhythmically against a table leg. Will drummed his fingers on the tabletop.

Xerxes continued, "Many years ago, the Council of Guardians resided on what is now GSS. The power sources were stolen during a raid by the Gallians and sent to a secret location. After much searching for the objects, the Weevil learned that the Gallians knew their whereabouts. The Weevil took over the space station, but didn't find the power sources."

Will stopped drumming his fingers and stared at Xerxes. "These power sources - they were used to power the *Leviathan*?"

"Not quite. Someone was contracted to build twenty Leviathan-class ships. The race of aliens that was hired for this operation developed a process where the strength of one of these power sources could be magnified many times. As a result, each of these sources was installed in one ship and equipped with this new system of power magnification."

Will inquired blandly, "You know all of those ships were destroyed, don't you?"

Xerxes grinned at him. "You don't realize how famous you became when you raided Ramses-3 and destroyed all those ships – but one."

"How do you know there is only one?" Jack asked.

"Word of your escapades carried far and wide across the universe. You set our enemies back so far that you became their number one target."

"Gee, it's great to be popular," Will kidded.

"Not really. They will stop at nothing to get you. Furey put together a whole alliance just to stop you. She even plotted to have you killed with your dad and Tera before all this started."

Will stared at Xerxes. "How do you know about my dad and Tera?"

"Tera was my daughter. We suspected that Furey and her renegades would ambush them. I tried to convince your dad not to go to your graduation but he refused to listen. He was determined to see you graduate from the academy. There was no way Tera would let him go alone. None of us suspected Furey was working with the Boromians and the Weevil at the time. She had everyone fooled." Will became visibly shaken by the news.

"So what does the *Leviathan* do for her?" Jack asked.

"This base was designed to control portals to and from quadrants all over the universe for peaceful purposes. Most of our power was channeled from the Guardians. On our own, we can only power one portal. At this time, we can only use our power to block the portal from GSS. They have tried for some time to overpower us."

"Is the portal the only way they can get to you?"

"No. They've created anarchy here on Earth. We are under constant siege by mercenaries and our defensive resources are dwindling."

"We're on Earth?" asked Will.

"Yes you are."

"Why did you bring me here?"

"Because, if we are to continue to fight this war, we need you to get us out of here."

"Who are we?"

"Me and the five hundred men and women stationed here."

"But, Xerxes, where would I put them all?"

"Do you have *Leviathan*?"

"What does that have to do with it?"

"We can control portals around the universe with the *Leviathan*. We need that to stop Furey and her allies."

"How long can your people hold out?"

"Not much longer. Our power bank is down to twenty percent capacity. At ten percent, we're defenseless."

"This isn't going to be easy."

"I know. Your father left you the device and the note because we both knew you'd be our last hope."

"But I'm just one person."

"I'll be your advisor and mentor. I owe it to your dad and Tera to look after you for them." Tears formed in Will's eyes. Xerxes explained sympathetically, "I know this is a lot for you to deal with, but there are many enemies out there waiting for an opportunity to get to you."

"Who are these enemies?"

"Oh, there are plenty of them."

Will slammed his fist on the table and demanded, "Who are these enemies?"

Xerxes was rattled by Will's directness. He replied, "You know about the Boromians."

"They aren't enemies anymore."

Xerxes was taken aback and said nervously, "What? But they killed Billy and Tera!"

"No they didn't. Asheroff witnessed the Fleet assassinating my dad and Tera."

"You're kidding!"

"No, I'm not. Asheroff confirmed it."

"Asheroff! He's the Boromian commander."

"And my new ally."

"Then the Attradeans will be hot on your trail."

Will rebutted, "Too late. Tenemon is dead and the Attradeans have been reduced to only a defensive role under Siphra. The Boromians in this sector are down to a small fighting unit. They will fight for me."

"The Gallians are formidable. They will be coming."

"If we stop Furey and capture or destroy GSS, they won't be going anywhere."

"How do you know that?"

"I've been to Gallia and met their leadership. We aren't fond of each other."

"But no one's been to Gallia and returned. Not even the Weevil," Xerxes replied with awe.

"I have."

Jack added, "We took a beating but, thanks to Will, I'm here. My close friend, Maya, wasn't so lucky. She was tortured and lies in a coma as we speak."

"Maya?"

"Yes, why?"

"The first time I saw Will as an infant, a young girl named Maya cared for him like a grown mother and her child."

Will replied, "Yes, that's the very same Maya."

"There are many other species out there that are part of Furey's plan."

Will informed him confidently, "We have an alliance of our own. Emperor Rethos and Empress Atilena of Urthos are behind us as well."

"Why would they do that?" asked Xerxes, baffled.

"Because I saved their kingdom and because they support our cause."

"What cause is that?"

"Have you heard of Yord?"

"Only tales of it. Once it was a great economic metropolis. When I was young, I dreamed of going there. Unfortunately, it was destroyed before I got a chance."

"Well, it's back. Yord is quickly returning into the mecca of commerce that once made it great."

"That's impossible! I heard that Tenemon took the Eye and the Seers who made Yord what it was."

"He did. I took them back."

"That can't be!"

Will stood up and walked around the table in front of Xerxes. "I married one of those Seers. In fact, that Seer and I rule Yord."

Xerxes stared into Will's eyes. He said with a glimmer of hope, "You're serious, aren't you?"

"Yes, I am."

"Then perhaps we do have a fighting chance."

Will explained, "I do have the *Leviathan* but I've kept her a secret until recently. Steele tricked us into helping her flush the Weevil out of Fleet headquarters. As a result, she knows about *Leviathan* and our secret base."

"I wondered how you came to know her as Steele."

"Come back with us to Yord and we'll devise a plan. Perhaps we won't have to move all these people at once."

Xerxes considered the proposal briefly and then replied, "Let me inform my people."

He picked up the phone and dialed three digits. "Steven Quentin, come to conference room Seven-Alpha."

Jack asked, "What kind of fighting force do you have?"

"We have the remnants of the United States military around us. Washington, D.C. and the immediate surrounding territories are all we control. Mercenaries and gangs rule the world."

"Wow. That's hard to believe. I was here on Earth as a kid and it was an awesome place to be."

"Not anymore."

A young man with brown hair and glasses entered the room. He wore a dark pinstripe suit with shiny black shoes.

"Xerxes! Are you alright?"

"I'm fine. Just relax, Steven."

Steven backed toward the phone. "Who are these men?"

"This is Will Saris, the one we've been waiting for."

Steven Quentin stared at Will and remarked, "You're the son of Bill Brock?" Will nodded. "I can't believe you finally got here. We'd just about given up."

Xerxes told Steven, "Things are much better than we thought. I'm going to meet with Will and his team. I'll be back shortly with a plan."

"But, Xerxes, what about the rest of us?"

"I won't be long. When I return, we'll be ready to take action, whether it's here or in another location."

"We won't give up on Earth. That's a promise," Will assured him. Steven didn't look totally convinced.

"The Attradeans and Boromians aren't an issue any more," Xerxes told Steven." Will claims that Rethos is his ally as well." Steven looked surprised. "And he has *Leviathan*," Xerxes added.

Steven jumped up and shouted, "We're saved. Hallelujah. Can I tell everyone?"

Xerxes looked to Will for a response. "Sure," Will replied cheerfully. "Why not? Everyone else in the universe seems to know."

"Are we ready?" Xerxes asked.

"The sooner the better."

Jack entered the portal first. Will extended his arm politely for Xerxes to proceed next. "After you, sir."

"Thank you, Will. Thank you very much."

When they emerged from the portal, Celine stood up and yelled, "Freeze or you're dead!"

Jack was startled and said, "Easy, girl. It's me." Xerxes came through the portal behind Jack.

Celine hollered, "Who's that behind you?"

Jack turned around and put a hand on Xerxes' shoulder. "This is Xerxes. He's a friend of Will's dad."

Will emerged from the portal and paused for a moment, looking at Celine. He was quite amused by her. Celine lowered her pistol and stowed it in the holster on her waist. "What's that look for, Will?"

Will laughed at her and said, "Gee, Celine. I never saw you so serious before."

"I'm glad you find it funny. Mariel chewed off three quarters of my right ass cheek because of you."

"Where is she now?"

"They took Maya up to Shanna's room and started their spooky stuff."

Will corrected her, "It's not spooky stuff. It's the power of healing."

"Well, I never healed like that."

"Why don't you fix the table before you fall over it," Will chided her. Celine righted the table, embarrassed.

Xerxes noticed the trunk with its open lid. He reached into it and retrieved the skull. "Do you know what this object can do?"

Will shrugged his shoulders. "I couldn't even guess."

"This skull has much power in it, but you have to know how to use it."

"Who would put power into a skull?"

"Your father took this from a powerful wizard. It's saved him more than once."

Will studied the skull then asked, "How does it work?"

"You can only use it defensively. It will repel any force used against you with exactly the same amount of force. But most importantly, never hold it in your bare hands when you use it. It could be fatal."

"Is there some method to make it work, a magic word or something?"

"No, you just hold it out in front of you like a force field. It will do the rest."

Mynx and Neva entered the *Phantom*. They were surprised to see Will and Jack with a stranger.

"Mariel just chased us down here to guard a portal you opened" Mynx complained. "I thought she was losing her mind."

"We're back already. It's okay," Will assured her.

"Then we'll leave you to your business."

"You might want to stay for this."

Neva complained impatiently, "I hope this is worthwhile. We're close to cracking the codes from GSS."

"Don't waste your time," Xerxes said. "Those are the codes the Guardians used to distract those who might attack the space station."

"What are you talking about?" Mynx asked him.

Will explained, "Xerxes' people built the space station that was taken over by the Gallians and then the Weevil. He knows all about it."

Neva replied sarcastically, "So we've been wasting our time?"

"I'm afraid so," Xerxes said.

The girls promptly sat down at the table. Mynx remarked, "I think we're going to find this more than a little interesting." The men sat as well.

"Do we need anyone else before you start?" Celine asked.

"Talia, Arasthmus and Bastille might want to be here," Will suggested. "I think Asheroff should be here, as well. Can you get them?"

"Sure," she said and then left.

Xerxes looked about the main quarters and commented, "If I didn't know better, I'd say this is an Attradean battlecruiser."

Will chuckled. "How right you are."

"It appears you lead quite an exciting life, Will."

"That's my goal."

"Just like your father."

Asheroff entered and joined them at the table. "How are the repairs to your ship coming?" Will asked him.

"Very well. The women you sent to help us are quite ingenious. I must admit, I was skeptical at first. I'm even more impressed by your underground dwelling. It seems you pulled one over on me."

"You understand, my friend, why it was necessary. Asheroff, this is Xerxes. He's a former member of the Council of Guardians."

"I understand you dissolved the Council, Xerxes," Asheroff said.

"Yes, I did. They became corrupt. I must admit, I'm absolutely astonished to have this conversation with you. I've heard all the stories about you and I never, ever imagined fighting on the same side."

"You can thank Will for that. He saved our ship, my life and the lives of my crew. The Weevil left us for dead."

"Well, I'm certainly happy we aren't opposing you anymore."

Celine entered with Talia, Bastille and Arasthmus. Will kidded, "This is the first time we've had this many people together in the main quarters of the *Phantom*."

"What's the purpose of this meeting, Will?" Arasthmus asked.

"Xerxes has joined us from Earth. We both have valuable assets that can be used to help each other in our war against the Weevil and the Gallians. We're here to discuss how we're going to use them."

Will felt something stir inside his head. It was Shanna playing with him. He tried to tell her telepathically, "Not now."

"I have to tell you - everything is fine. Maya is responding well. She's going to be fine." Will looked at Jack with a smile. Jack smiled back and gave him a subtle thumbs up. "Thanks, Shanna. I have to focus. We'll talk soon." She didn't reply. Will cringed because he knew what that meant.

"Are you alright, Will?" Xerxes asked. "You look distracted."

"I just got some good news."

"How? I didn't get it."

"It was cloaked."

"You can do that?"

"Uh-huh."

"Boy, I have missed a lot."

Will stood up and walked around the table. "Here's what we know. If anyone disagrees, please speak up and let me know what's wrong. We know that the Weevil control GSS, the space station controlling portals all over the universe." He looked around the table at everyone but no one disagreed.

"Steele Furey is leading the Gallians and using the Fleet to accomplish her bidding. She now requires the *Leviathan* to complete the next phase of her plan." He paused again, waiting for objections but none came.

"The Weevil are trying to conquer the transport station on Earth, the last obstacle preventing them from controlling the planet." Will looked to Xerxes and asked, "Why are they using human mercenaries instead of their own soldiers?"

"We suspect that the major part of their invasion force isn't here yet. It would be much easier to conquer a planet in total chaos with no real ruling power."

"Any comments?" Will asked everyone.

Xerxes added, "We know that the Weevil have a much larger force in the fifth quadrant. It would take light years to get them here unless they can use one hundred percent of their power to establish twelve portals necessary for the force to pass through."

"Why can't they reach one hundred percent power?" Will asked.

"Because we are countering twenty percent of it in defending Earth,"

"Any other factors?"

"What about the *Leviathan*?" Xerxes asked. "If the Weevil get their hands on it, they could create those portals on the ship's power source alone. They wouldn't need the space station anymore."

Bastille inquired, "This Steele Furey is a tactical commander. Why hasn't she tried to retake the space station? Are the Weevil that powerful?"

"I wondered that, myself," Asheroff said. "The Gallians didn't put up much of a fight to recover the station. If they are better prepared for battle, what are they waiting for?"

Maya's voice from the hatch startled them. "Because she wants *Leviathan* to control all the portals."

Maya stood with the aid of a crutch next to Shanna and Mariel. Bandages covered the wound and bare scalp on the left side of her head. The other women wore white robes, appearing more like angels than battle-worn survivors.

Jack jumped up and blurted, "Maya!"

Maya continued, "She'll make everyone pay to use the portals, including the Gallians. If she had the *Leviathan*, no one would ever be able to stop her."

Will introduced her to Xerxes. "I remember you from when I was a child," Maya remarked.

"And I, you."

"So all this treachery among races breaks down to the simple acts of blackmail and greed," Asheroff concluded.

Will replied, "Pure and simple. What if the Gallians and the Weevil are really on the same side? Who is keeping the portals open in route to Gallia? The Gallians don't have the power anymore."

Asheroff placed his hands on the table and stood up. "She's using everyone in the sector!" he declared angrily. "That becomes all so clear now."

"I trained under Steele. I know her," Maya told him. "I know how she thinks and how she acts. I always thought she limited her activities to those that were honorable."

"It appears that we now know her motive. If we deny her the *Leviathan*, we also have to deny her the transport station on Earth," Will said.

"We need to put the *Leviathan* where she won't expect it," Asheroff suggested. "Then we have to figure out how to use it from that location."

"I've got a suggestion," Jack said. "If the *Leviathan* has all this power, why not use it to overpower the space station from the transport center? Why not put the *Leviathan* inside the transport bay and offset more than twenty percent of the intruding force."

Will countered, "We can do that, but we have to support it with something else. We can't let Steele focus all her resources on Earth."

"What if we baited her into a meeting on Gallia?" Shanna suggested.

"For what purpose?" asked Will.

"If we can shut down the portals with her on the Gallian side, she's isolated from the Weevil."

"Then I can take control of the Fleet," Maya added.

"What would you do with them?" Jack asked.

"With Steele out of their range, I can launch an assault on the space station. When we have everything on this side under control, we'll open the portal and go after the Gallians."

"If we're there, we can negotiate a treaty with them," Shanna continued. "They'll see Steele's power crumbling around them."

"I don't like it. It's too risky," Will said.

Xerxes reminded him, "You have other options available to you."

"What if we showed one ship from each ally representative of our strength to the Gallians? If they saw how strong we are, they might buckle," Asheroff suggested.

"That's good," Will told him. "I'm sure we can get Rethos to provide a ship and Siphra as well."

"Do we want the *Leviathan* there?" Bastille asked.

"Absolutely not. Once the space station is shut down, I want the *Leviathan* available to help restore order on Earth."

"Who will go down to the Gallian surface to negotiate?" Arasthmus asked.

Xerxes telepathically spoke to Will, "Remember the skull. They can't harm you." Will recalled and smiled.

"I will. I'll go there and negotiate," he said.

"I'll go, too." Shanna told Will.

Will noticed Mariel was impassive. "What do you think, Mariel?" he asked.

"This skull, you speak of…"

Will replied, "It's more of a secret weapon, I guess you could say."

She walked to the center of the main chamber and looked at the faces of everyone at the table. "I will go, too. I don't think this will end the war with the Weevil but if we can cut them off from our sector, at least for a while, they may think twice before coming back."

"I don't think it's a good idea for any of you to go with me. I'd rather do this alone," Will said.

Shanna approached him. "We've been through this before, Will. We're going and that's it."

He looked at Jack. "Perhaps you should stay with Maya," he suggested. "She'll need help without Zira to assist her."

"Thanks, Will. I'd like that."

"What will we do with Furey when we have her?" Asheroff asked.

"It will be sweet and it will be fitting," Maya promised.

"Maybe she should receive the Fleet's method of execution for a traitor," Will offered.

Maya grinned. "Oh, no. That would be too kind. I have something much better in mind."

"Can you take your ship and accompany us to Gallia?" Will asked Asheroff.

"I'd be happy to. If you need ground assistance, we'll be there for you as well."

"Thanks, Asheroff." Will looked around the table and then asked, "Did we cover everything?" Everyone nodded in agreement.

"Then we'll begin our operation shortly. I want to coordinate the times and arrivals of our ships with each of you, beginning the day after tomorrow. Until then, everyone get some rest. We have a lot of work ahead of us."

"I'll return and coordinate my forces to support the *Leviathan*," Xerxes said.

"Are you confident in the plan?" Will asked Xerxes.

"Of course. I can't believe our day of reckoning has finally come. Maybe we have a chance for peace after all." He shook Will's hand and left the main chamber through the portal.

Asheroff also shook hands with Will and remarked, "You have a fine team around you. No wonder you came out on top."

"You're part of this team now, too." Asheroff cracked an odd smile and left.

Arasthmus and Mariel approached Will. "Perhaps I underestimated you, Will," Mariel admitted.

"Thanks, Mariel. Thanks for everything. Obviously what you did for Maya worked quite well."

"She's a strong woman. I was worried when things happened quickly between us but Shanna was the stabilizing force."

Will pulled Shanna next to him and confessed, "Shanna is always a stabilizing force." Arasthmus and Mariel smiled and left them.

"I think you should let Neva and me participate in the raid on the space station," Mynx suggested. "We do know that part of the sector pretty well."

"If I do that, Neva will beat the snot out of every Weevil on board," Will joked.

"Not if I get them first. I only need one arm and a knife to fillet those bastards."

"You're in. I'm sure Maya will be glad to have you." The girls beamed with satisfaction as they left the ship.

Will noticed the portal shut down. He realized that Xerxes could do that as well.

"How are you feeling, Maya?" Will asked.

"I'm getting stronger."

"Do you recall what happened on Gallia?"

"Even better. I was inside Steele's mind. I knew her thoughts and feelings. I haunted her and she couldn't do anything about it. Unfortunately, I got stuck there, probably from the shocks."

"Well, we all missed you very much."

"Would you mind if we retired for the evening?" Jack asked. "I haven't slept in a while and there are a few things I have to talk with Maya about."

"Sure."

"Take it easy on him, Maya," Shanna warned

Maya replied playfully, "He's got a one night reprieve. After that, he's at my mercy again."

Jack and Maya left the *Phantom* and returned to the *Luna C.*

Will asked Shanna, "Where's Marina?"

"Keira and Regent are watching her."

"Let's go see our daughter."

Shanna grabbed Will by the collar with both hands. "Not until you kiss me." Shanna stared at Will with puppy dog eyes.

"Of course, dear." Will embraced her and kissed her fervently.

XI

THE TRAP

W ill and Shanna left the secrecy of the underground base and went above ground to the new palace. Keira sat inside one of the bedrooms, gently rocking their baby. Shanna and Will entered, pleased to see them. Shanna asked Keira, "How is she doing?"

"She's eating and drinking just fine."

"Thank you so much for your help."

"You're welcome." Keira handed Marina to Shanna and offered, "Anytime you need me, just holler."

"We're really grateful for everything you've done, Keira," Will added. "With so much going on, it's been difficult to be here for her."

"I was glad to be of service."

"You'd best get going. I'm sure Regent is missing you," Shanna told her.

"I'm sure he is, too. Good night," she bade them and departed.

Shanna held Marina out for Will to see. "Mind if I hold her a bit?" he asked.

"Of course not. You're her daddy."

"How are you feeling?" he asked.

"Sore. I didn't expect giving birth to be so stressful although Keira said that it was actually a smooth one, mostly because of half-pint's size."

"Did she say it's okay for you to be on your feet already?"

"I just have to be careful about lifting anything heavy or running up the steps for another day or so."

"You'd better be careful. I don't want anything to happen to you."

Shanna nestled her head against his shoulder. "You do care about me, don't you?"

"Of course, I do. Shall we sit on the balcony?"

"Sure. I could use some fresh air."

Will pushed open the tall, stained-glass doors and stepped outside. Shanna picked up a shawl from the rocking chair and placed it over her shoulders, while grabbing a small blanket for Marina. She joined Will on a swinging bench. They looked out at the main street in the distance. Shanna commented, "Look at the beautiful lights down there. It's so colorful."

"I can't remember the last time I saw so much activity at night. It seems like everyone's out tonight," he replied. Will placed his thumb in Marina's little palm. She squeezed it tightly and closed her eyes.

"You know, Shanna, parenthood makes all our problems seem so far away."

"Yes, it does," she replied and then paused for a moment. "How are we going to pull off the next stage of our plans, logistically speaking?"

"I think by sending the *Leviathan* to Earth the day after tomorrow, we'll throw a real monkey wrench into Steele's plans."

"What do you think of Xerxes? Can we trust him?"

"I'm sure we can. I'd like to spend more time with him, but he's got his hands full at the transport center on Earth."

"I'd like to go to Earth sometime. I've heard a lot about it."

"Maya could tell you quite a bit. She's picked up a few interesting habits from there. Xerxes did say that it was engulfed in anarchy, though."

"How will we find Steele?" she inquired.

"We're going to let the Gallians find her for us."

"So we're going back to Gallia then."

"That's right. I have a surprise for them."

"Then I'll be by your side."

"I'm sure you will."

"What's that supposed to mean?" Shanna asked, defensive.

Will laughed. "I'm sure you won't let me out of your sight from now on, until this is over."

"That's right. Is there something wrong with that?"

"No. I like it."

Shanna wrapped her arms around Will and the baby. She leaned gently against his side. Will gazed at them and smiled briefly before turning serious.

"We have to get Steele on Gallia before Maya attempts to gain control of the Fleet. Once that happens, I'm hoping they can take control of GSS. With the Fleet supporting her, that could happen quickly."

"How do you plan to deal with Steele?"

"Maya wants a piece of her in a bad way. I'd be happy if we could coax her into a treaty of some sort."

"She won't give up. You know that, don't you?"

"Uh-huh. I've been thinking about that."

"I don't understand why the Gallians call her 'Cherenka'. Why didn't she stick with Steele?" asked Shanna.

"She's not a normal woman by any means."

"Maybe she's not a woman at all. Maybe she's a cyborg."

Will thought about that for a moment. "That's an interesting theory. Steele Furey - Steel Fury. Jack told me she was awful smooth during the raid on the Ceratoan supply station. He also said she was a sniper with the pulse pistol."

"So how do we stop her?"

"I don't know. Maybe we're jumping to conclusions about her. But then again, I'd hate to underestimate her." Will yawned and leaned his head back. He gazed at the stars and wondered if there would ever be peace in the galaxy.

Shanna sat upright and took Marina from him. "I'll tuck her into bed. Why don't you come as well? It's getting late."

"I'll be there in a bit."

Will stood and scanned the tops of the many buildings. He watched the twinkling of several lights in the distance. An eerie chill made his spine tingle as he thought, *Steele, I know you're out there somewhere.*

Will paused in the doorway and sensed something dangerous about to happen. He saw a red dot dance off the mirror onto the wall. "Shanna, get down, quickly," he whispered.

"What is it, Will?"

"Get down, damn it!" Will stepped away from the doorway and grabbed his pulse pistol. He searched the rooftops across the street. "Someone's targeting us." Shanna huddled on the floor and waited.

"She's out there. I can feel her." He looked further away at the next tier of buildings.

Shanna muttered in frustration, "She could be anywhere."

Will pushed the door shut and pulled the curtain across. "Leave the lights off. We have to get down to the base." Shanna scooped up Marina and followed Will out of the room.

· · · · · · · ● · · · · · · · · ·

They reached the dock and paused between the *Luna C* and the *Phantom*. Will instructed Shanna to take Marina on board the *Phantom*.

Jack and Maya emerged from the *Luna C* in a hurry. Maya remarked irately, "You felt her, too, Will."

"Yes, I did. She's on the surface."

"I think it's time to play."

"I think so, too. Get Talia, Mynx and Neva. Take the *Luna C* and get into position as soon as possible. Don't do anything until you hear from me."

"You and Shanna are still going to Gallia?"

"Yes, we are."

"Be careful," Maya warned and hugged them.

Jack shook Will's hand and then he, too, hugged Shanna. "You both be careful."

"Don't worry," Will assured him, "I've got a surprise for them."

"What do you want me to do?" Shanna asked.

"Go to Mariel and explain what's happening. Leave Marina with her. I've got to talk to Bastille and Celine." Shanna hurried from the ship and ascended the stairs.

Will boarded the *Leviathan* and climbed the stairs to the second level. He knocked sharply on Bastille's door. "Who is it?" Bastille's gruff voice bellowed,

"It's Will. We have a situation." He heard bumping and banging from inside Bastille's room.

Finally the door opened. Bastille stood awkwardly with a blanket wrapped around him. His hair stood almost straight up. "What's going on?"

"Steele's on the surface nearby. I don't know if she's alone."

"What do you need me to do?"

"Get your group together and get the *Leviathan* out of here now. You're going to Earth. I want you to find the area known as Washington, District of Columbia. When you get close, a man named Xerxes will contact you with landing instructions."

"We'll be out of here within the hour."

"Thanks, Bastille. Have Xerxes contact me when the *Leviathan* encounters GSS's forces. Good luck and keep everyone safe."

"You, too, friend."

Will returned to the *Phantom*, where Shanna waited inside the hatch. She wore her black leather battle garb, including the knives in the arm and leg straps. Draped from her belt were four trains of sheer black material, covering the front and back of her legs. She wore a pulse pistol on her belt and held two of her knives in an attack posture.

Will marched through the hatch and froze, amazed at how good Shanna looked in the outfit. "Does it still fit me okay?" She asked nervously.

"Uh, yeah. You sure don't look like a mother."

"Why thank you, Will."

"How did it go?"

"Everyone's on the move. As soon as the *Leviathan* and *Luna C* pull out, we will too."

Will pressed the knob and closed the hatch. He turned off the hatch power switch to ensure no one entered from outside. "Steele doesn't know how to get down here from the surface, so that should buy us some time."

"Would you mind if I got some rest?"

"Go ahead. I'll wake you if I need you."

Shanna ascended the stairs to the flight deck. She sat in the co-pilot's seat and quickly fell asleep. Her body hadn't replenished its strength yet and she was still sore from Marina's birth.

Will paced the main quarters wondering if he was doing the right thing. For some reason, he feared confronting Steele. When he entered the flight deck, he gazed at Shanna and admired her as she slept. He recalled how his life had changed since he left the academy. So many things happened between him and Shanna since that day he rescued her from Tenemon's prison. Was it destiny? Fate? Or maybe it was all just coincidence.

Will checked the scanners and saw a clear monitor. He thought, *What about the people of Yord? Who would have thought so many of them were of Firenghian blood? They're my people.* The big question Will asked himself, *Is my destiny already decided?*

Bastille's voice pealed from the speaker over Will's head. "Will, are you there? It's Bastille?"

"I'm here. Go ahead."

"Everyone's on board and we're pulling out."

"Excellent."

"There's a small problem, though."

"What is it?"

"I can't raise Xerxes or any of his people. I used the coordinates and frequencies he left but there's no response."

Will tensed as he realized what might be happening. "Continue on, Bastille. Shanna and I are going to Earth via the portal. I'll see you there."

Will watched the *Leviathan* on the monitor as it pulled out of the dock. He was proud of the ship, yet he feared what she was capable of. In the wrong hands, it could be disastrous.

The ship was the size of a small town, easily able to house two thousand people. Its weapons arsenal was advanced beyond anything else ever developed. Its power supply gave it the capability to make its own portals to any location in the universe. It was truly a miracle of engineering by whoever designed it. Fortunately, there was only one *Leviathan*. The titan ship vanished down the tunnel, leading from the bay.

Will leaned over and kissed Shanna's cheek, then leaned back and closed his eyes. Three hours later, Maya's voice woke him from a deep sleep. "Will, it's Maya. Do you read me?"

Will rubbed his eyes as Maya's face appeared on the monitor. "Maya, what's up?"

"I have the commanders of five Fleet ships supporting me. They're aware of Steele's plot and are willing to fight against her."

"You're sure they're not Weevil?"

"They passed the test."

"That's great. How about the others?"

"Some are in on the plot with her. I've been advised which ones to avoid."

"What do you recommend?"

"Give me the sign and we can attack GSS with what we have."

"Let me know when you are in place. Shanna and I are going to Earth. Xerxes isn't answering his calls."

"Damn!" she uttered. "That's not good."

"No it isn't. We're using the portal to get there. I'll call you as soon as we assess the situation."

"Be careful, Will. It could be a trap."

"I'm sure it is. You be careful as well. Watch your back and Jack's, too."

Jack's face appeared next to Maya's on the monitor. He interrupted them kiddingly, "Well, it's about time you acknowledged me, pal."

"I miss you, too," Will kidded.

"Eve's for drinks?"

"Definitely, for a victory toast."

Maya cut in, "That's enough, boys. We have work to do."

"I'll contact you when we find out what's happening on Earth."

"See ya'," Maya chirped. The monitor went blank.

Shanna stretched and yawned. "Well, hello Sleeping Beauty," Will teased.

"Hi, yourself. Are we there yet?"

"Change of plan. We're taking a portal to Earth."

Surprised, Shanna asked, "What for?"

"Xerxes isn't responding to Bastille's calls. Something's up."

"You think Steele got there already?"

"I don't know how?"

They descended the stairs to the main quarters. Will asked with an air of concern, "Are you alright for this?"

Shanna replied confidently, "I'll be fine."

Will reflected a moment. "Did you ever wonder how we keep dodging bullets like this?"

Shanna replied confidently, "It's fate. We have a destiny to fulfill."

"I hope you're right." He retrieved the portal control device from the trunk and strapped it on his left wrist. He opened the black box and removed the brown sack with the skull in it.

"What's the deal with that?" she asked, curious.

"Do you know what an ace up the sleeve is?"

"Not really."

"It's a backup plan."

"Oh, okay."

Will latched the sack's drawstring to the front of his belt. Next he strapped a pulse pistol and holster to his left side. Shanna helped tighten the belt. "Now you look like you're ready for business."

"Not yet, I don't." He fastened his sword and scabbard around his waist and positioned it on his right side. He placed an arm around Shanna's waist and kissed her lips. "Now, I'm ready for business." Shanna became excited but pushed him away.

"I can't believe you just did that," Will teased.

"Not now. I'm not ready yet," Shanna said shyly.

"What's this 'not ready' stuff?" Will laughed.

Shanna was embarrassed and replied defensively, "I told you, I need a few more days. You know, having a baby does that to you."

"I know. I just enjoyed being the aggressor for once."

"Don't worry. It'll be the last time," she warned.

Will aimed the portal control device at the wall and pressed three buttons. The blue circle again formed and became blurry. "Ready?" he asked.

Shanna gazed at the circle and inquired, "What is it?"

"It's a portal. Come on, I'll show you." He took her by the hand and led her into it. Shanna was apprehensive as they walked through teal clouds and followed a narrow path of blue sand. She leaned over the side and looked down.

Will bumped her and kidded, "Don't fall."

Shanna nearly lost her balance and grabbed Will's arm. "You jerk! What if I fell?" she chastised him.

"I wouldn't let you fall."

Shanna looked down again into the mist and wondered aloud, "What's down there?"

"I don't know," replied Will, also mesmerized by the colors and sands. "I can't tell if there is a *down there*."

They emerged from the portal and stepped into the conference room. Sensing danger, Will drew his pistol and hurried to the door. He looked back at the portal and aimed the device at it. He pressed a red button and the portal closed. "Not bad, huh?"

"Okay, Mr. Wizard," chided Shanna. "So you did something right."

Will grinned and peeked down the hallway. Shanna drew two knives and stood behind him. She whispered, "See anything?"

"Not yet." Will stepped into the hallway and crept toward the elevator. He motioned for Shanna to halt. She raised both hands and poised to throw the knives when she acquired a target. Will stood near the corner in front of the elevator. He heard the sound of soldiers' footsteps marching toward them.

Shanna waited anxiously for directions. Will showed her three fingers and pointed toward the elevator. Shanna nodded. Will stepped toward the elevator, ignoring the approaching soldiers. He pressed the down button.

Three Gallian soldiers rushed toward him. "Stop right there," one of them ordered. Will stowed his pistol and put his hands up. Shanna crept closer. When the soldiers surrounded Will, one of them asked, "Who are you and what are you doing here?"

Will replied cynically, "I was going for my physical and got lost."

Shanna fired both knives, striking two of the soldiers in the sides of their necks. The third was surprised and turned to fire his gun. Will whirled around and rammed the soldier with his shoulder. They tumbled to the ground, wrestling wildly.

Shanna drew another knife and grabbed the soldier by the neck. She held the blade snug against his jugular. "Game over, pal. Don't move." The soldier reluctantly released his hold on Will and put his arms down.

"I could have handled him," Will complained.

"Come on, Will. He was whipping your ass," Shanna taunted him.

"No way!" Will looked the Gallian in the eye and asked, "Were you whipping my ass?" The soldier remained silent.

Shanna encouraged him, "You can answer honestly. We won't hold it against you."

The soldier answered apprehensively, "Yeah, I was whooping your ass."

Will stood and pretended to be angry. "This really sucks. Every time I get in a fight, I get my ass whooped."

"Are you going to kill me?" the Gallian asked.

"Today could be your lucky day. Don't ruin it."

Shanna took Will's pulse pistol and aimed it at the Gallian. Will helped him to his feet. "All I want from you is a little information."

"You know I can't do that."

"What if I could convince you?"

"How? Torture?"

"No, by information exchange."

The soldier laughed at him and remarked, "You're funny, human."

Will became annoyed. "Look at my eyes, numb nuts. Do I look human?"

The soldier noticed Will's feline eyes and, growing bored, he inquired, "What's your point?" Will ignored him and waited impatiently.

The elevator bell chimed once and the doors opened smoothly. There was no one inside. "Get in the elevator, now," Will ordered. The soldier promptly obeyed.

"Shanna, block the door open and keep an eye out. I don't want the elevator to move unless we need it to."

Shanna retrieved her knives from the dead soldiers and stood in the doorway. She watched the corridors alertly.

Will explained to the soldier, "Here's the problem. Steele, I mean Cherenka, is setting up your people. You guys lost your space station and your ability to run an intergalactic shipping conglomerate, right?"

"We were a peaceful people. The Weevil started all of this."

"What if I told you that you can have GSS back? You can rebuild your shipping business."

"Why would you let us do that?"

"Because I'm trying to make things right in the sector again. Yord is being restored as we speak. I'm sure it could profit by having a shipping group like the one you once had to send and receive goods from around the universe."

The soldier snorted and commented, "The Major would never go for that."

"What if I told you that Cherenka is going to sell you out to the Weevil?"

"She would never do that."

"She would and she is. She's going to turn on everyone."

"What good would that do her?"

"I see you're catching on. She wants the *Leviathan* and she's using the Gallians and Weevil to provide a large enough distraction so she can hijack it."

"Who's got this *Leviathan* I've heard so much about?"

"It doesn't matter right now. When she gets her hands on it, she can control portals with it. She'll have enough firepower at her disposal to blackmail everyone and anyone. She'd be unstoppable."

"So what do you want from me?"

"Tell me what happened here?" The soldier refused to answer.

Will warned, "We can do this the easy way or the quick way." The soldier still refused to answer.

Will nodded to Shanna. She placed the gun to his temple.

"Alright! Twenty of us hid inside a freighter. When the freighter landed in the transport bay, we came out and attacked."

"Where are your friends?"

"Some were killed. The rest of us were forced up the stairwell. We split up on different floors."

"Do you think you could convince your peers to help us stop Cherenka?"

"I don't know."

Will motioned for Shanna to get on the elevator. She quickly ducked inside and paused by the control panel.

"Where to?" she asked.

"Bottom level. Transport bay."

The soldier became nervous. He fidgeted and moved toward the side of the elevator. Will asked, "What's wrong?"

"They'll kill me as soon as they see me."

Will mused, "Good point." He and Shanna stood on the opposite side, facing the man.

Will cautioned him, "Don't move and you'll be alright."

The elevator lurched to a stop. The bell chimed and the doors slid open. They could hear feet scurrying outside the doors. "Hold your fire," Will shouted. "We're friends."

Someone yelled, "Drop your weapons and come out, now."

Will replied, "Where is Xerxes? I need to speak with him."

"Drop your weapons and come out!"

"I'm going to tell you one more time; let me speak to Xerxes," Will declared. The soldier fired a shot into the back of the elevator.

"I hate dealing with idiots," Will whined disgustedly. He pressed the button for the next level.

Before the elevator's doors closed, the soldiers fired numerous rounds into the back of the elevator wall. The sound was ear-shattering as shards of plastic and lightweight metal sprayed all over the floor. The door finally closed and the elevator rose to the next floor.

"I see you handled that well," Shanna goaded him playfully.

"Oh, bite me, Shanna. You try next time."

"Maybe all it takes is a woman's touch."

The soldier drew a deep breath and muttered, "Oh, joy."

Shanna yelled at him, "Did I ask for your opinion?"

Will chastised her, "Now wait a minute. You didn't mind it when he took your side."

The soldier asked, "Can I say something?"

"What is it?" Will asked.

"I think both of you are crazy."

"Did you ever see people turn into animals?"

The Gallian snickered and replied, "Come on. You expect me to believe that one? Now I know you're insane." Will stopped the elevator and lowered his head. The soldier laughed at him and teased, "You two are pretty funny. This is entertaining." Will lifted his head. He had the face of a wolverine.

The soldier shuddered and froze against the wall. "Alright! I believe you!" He looked shaken, fearing they were supernatural. Will looked away until his face returned to normal.

Shanna kidded, "Would you like to see me change, too?"

"Please, no! I've seen enough," the soldier pleaded.

"Is there anything else you'd like us to prove to you?" Will asked.

"No! Absolutely not! I believe you."

"Good. Now we can continue our mission."

Will pulled out the stop button. The elevator jerked slightly and ascended to the next level. Will warned, "Get ready." The doors opened slowly.

Shanna shouted, "Don't shoot. Is Xerxes out there?"

A woman's voice yelled back, "Who wants to know?"

"I'm a friend of his. He's expecting me."

"It sounds like you know who." Will whispered.

Shanna nodded. "Give him a message for me," she yelled.

"What is it?"

"Tell him we'll call him from Gallia. He'll know what it's about."

"Like hell you will," the woman screamed. She opened fire at the elevator with a pulse rifle.

Shanna and the soldier covered their ears. Will cringed and leaned into the front corner for safety. He pressed the elevator's close button and laughed at Shanna. "Nice touch."

"Shut up, Will!"

"Alright, enough games. Now we get serious." Will took the pistol from Shanna and pressed the elevator's open button. The doors slid open. Will pressed the close button and dove out of the elevator.

Shanna called out, "Will!"

Will rolled across the floor and scrambled around the corner. "Cherenka, let's talk," he shouted. A moment of silence passed. "Cherenka, I know you're out there."

"Who wants to know?"

"I'm just a messenger."

"What is your message?"

"Your time is running out." He peeked around the corner and saw three Fleet officers with Steele. He heard one of them ask her, "Why does he call you Cherenka?"

Steele advised the officer sternly, "Don't you worry about it."

"Do you really think you can fool everyone like this?" Will asked her.

She recognized his voice and said, "It's gone on this long. Why shouldn't it keep working?"

Will used his telepathy to communicate with Shanna. "Do you hear me Shanna?"

"Yes, I do, Will."

"Can you hear Steele talking?"

"Real clear."

"Tell our friend to keep listening."

"He's interested already."

Will focused his attention on Steele again. "I'll bet you that the Gallians are already onto your little scheme."

"Hardly. They couldn't find their way out of a simple box without my help."

"Don't you feel the least bit guilty about it?"

"Come on, Will. You know me well enough, by now. I'm cold-hearted and selfish."

Will realized he had to change tactics since she had figured out who he was. He focused on Shanna again. "Are you there Shanna?"

"Yes, Will."

"Follow my thoughts. I'm going to move fast."

Will taunted Steele, "It's a shame you'll never get your hands on the *Leviathan*. She's headed to Gallia."

"Come on, Will. Do you really expect me to believe that?" Steele responded sarcastically.

"Believe this, Steele. Emperor Rethos, Queen Siphra and Asheroff are sending ships as well. The Gallians will negotiate a peaceful settlement with us. Then we'll expose your plan."

Will instructed Shanna telepathically, "When I tell you, press the open button and then immediately press close. I'm coming through in a hurry."

"I'll be ready."

Steele asked, "Why in the world would they help you, Will?"

"Because I can offer them a lot more in peace than you can in war."

"They're just pawns anyway," Steele mocked.

Will peeked around the corner and saw only Steele standing there. He knew other officers were coming around behind him. "Steele, there's just one thing I've got to know."

"And what's that, your Highness?"

Will instructed Shanna, "Now!" He dove across the floor and fired three quick shots at Steele's right knee. Sparks flew from her knee and thigh. The third struck her right shoulder and she fell to the ground. She shrieked and screamed, "Saris, you bastard! I'll get you."

The elevator doors opened. Will scrambled to his knees and lunged inside. "Just what I thought," he taunted. "You're a heartless pile of junk." He got up and pressed the 'TB' button to return to the transport bay.

"Why are we going back there?" Shanna asked.

"We're changing the plan. Work with me on this." Will breathed heavily as he tried to catch his breath.

Shanna teased, "What's the matter? A little out of shape?"

Will smiled at her and replied, "No, your suggestion that she might be a cyborg was pretty astute. Who would have thought?"

The Gallian soldier said humbly, "I owe you an apology. You were right."

"We're not out of this yet," Will warned. He handed his pistol to Shanna as the elevator doors opened. They heard the scurrying of feet again outside.

"I'm unarmed and I'm coming out," Will announced.

A man's voice ordered, "Place your hands on the back of your head and walk slowly."

Will complied and stepped out of the elevator. "Where's Xerxes? It's a matter of utmost importance."

Several soldiers encircled him and pointed guns at him. An officer approached Will and asked, "What do you want with him?"

"If you tell him that Will is here, I'm sure he'd be grateful."

"We're in a lockdown until we secure the facility. How did you get in here?"

"Xerxes brought me here." The officer laughed at him.

Will became annoyed and asked, "Are all officers this stupid or is it something in the air these days?" The officer punched Will in the face.

Shanna focused on Will and asked, "Do you want us to help you?"

"No. Close the door and stay hidden," he directed her.

Will said angrily. "I'm the one chance you have to be rescued and you just struck me."

The officer laughed again and said, "Oh, a smart ass, huh."

Xerxes approached and yelled, "What the hell are you doing?"

The officer turned to him and replied, "It's just some smart-allick punk."

"Take Colonel Winston's firearms from him and cuff him," Xerxes ordered the soldiers.

"You can't do this to me!"

Xerxes punched him in the mouth, sending teeth and blood flying across the floor.

Will chuckled and quipped, "Now that must have hurt." He asked Xerxes, "What's going on here? The *Leviathan* is on the way and they can't contact you."

"Our power is fading faster than we thought, Steele and some of her troops have come through on two levels from two portals they opened."

"How many people are here in the facility?"

"Over five hundred on the secured floors. Three of the levels above us have been breached."

"I have an idea." Will called to Shanna, "Come out with our Gallian friend."

The elevator doors opened again. Shanna stepped out with the Gallian soldier. Xerxes' men trained their weapons on the Gallian.

Will hollered, "No! Don't shoot."

"What do you have in mind?" Xerxes asked.

"Steele is on the floor above us. I coaxed the truth out of her and our Gallian friend heard it."

"So what can we do?"

"I've sent Maya and her force against GSS. Once we have control, we'll shut down the portal. Then we'll return and hunt down the rats."

"I'll arrange for my men to take a defensive stand," Xerxes responded.

Will asked the Gallian, "Can you convince your troops to surrender and return with us."

"How would I do that?"

"We have a page system," Xerxes suggested. "Inform them of the change in the situation and have them meet you in the stairwell."

Xerxes took a handset from a page unit on the wall and held it out to the Gallian.

Will informed the Gallian, "This is the future of your people. You have to choose one of two paths: freedom or slavery."

The Gallian took the handset and spoke in a garbled language for four minutes.

Shanna suggested to Will, "I think we should look for Steele's portal. When we find it, I'm sure we'll find her."

"You're one step ahead of me. Let's do it."

"I think it's about time we put her on the defensive," commented Shanna.

"Maybe she's not used to playing defense. That could be the advantage we've been looking for," Will suggested.

Will and Shanna ascended the stairwell to the next level. Ten soldiers immediately surrounded them but a lieutenant ordered them back.

"Mr. Saris, I'm Lt. Dixon. Xerxes ordered me to support you as required."

"Where did the insurgents enter from?"

"We believe they came from two locations. One is from the rear of the transport bay. The second was on the level above us. We secured the stairwells and programmed the elevator so it only accesses three levels. We control the upper and lower levels so they are trapped in the middle."

"We're going after them."

"Do you want any backup?"

"No, thanks. Just keep a sharp eye out. Steele's a cyborg and she's very dangerous."

"I understand."

Will and Shanna returned to the elevator. Will stared quietly at the wall while they waited. "You have a plan in mind?" Shanna asked.

"Yup. When the door opens, I'm going to leap out onto the floor. You lay down in here and face straight ahead, I'll face right."

"What if there are too many of them?"

"Then follow me. We'll shoot our way down the right hallway."

The elevator door opened and they stepped in. As the door closed, Shanna pulled Will to her and kissed him. He was pleasantly surprised. "What was that for?"

"For the fun of it."

"You're a real sicko, you know that?"

"Just for you," Shanna teased.

Will shook his head in disbelief. "What kind of a creature have I married?"

They each drew a pulse pistol and flattened against the left wall of the elevator when the doors opened. Will held up one finger, two fingers, and then three. He leaped out of the elevator and landed flat on the floor. There was no one in the right hallway.

Shanna laid on the floor of the elevator, facing out. She saw three Fleet officers enter the hallway ahead of them. She crawled out of the elevator and fired several shots. Two fell to the ground critically injured. The third ducked inside a doorway and returned fire. Will grabbed Shanna's ankles and pulled her behind a partition.

"Ouch!" Now I've got a rug burn on my stomach," she complained.

"Well, next time, wear some clothes."

Shanna rubbed her bare stomach and groaned, "Smart ass."

She poked her head around the corner and fired another shot, catching the third Fleet officer in the face with a blast from her pulse pistol. He fell to the floor and writhed in pain, desperately covering his bloody face with both hands.

"They won't be bothering us," Shanna remarked sarcastically.

They hurried down the hall. One by one, they opened doors and searched each room. "Maybe she left already," Shanna suggested.

"No. She's still here. I can feel her."

They rounded a corner and proceeded down another hallway. Voices came from an office halfway down the hall. Will motioned for silence as they crept closer to the room.

Inside, Steele lay on a desktop on her back. Two men were by her side. One of them wrapped her knee and thigh with a bandage. "Look, Steele, you've got to take it easy. There's damage to the tri-axial cords which provide multi-lateral movement of the knee."

"I can still walk. Can't I?"

"Yes, you can. But if you strain it in the wrong direction, you could snap the remaining cord."

"I'll kill that bastard, Saris. He got one lucky shot and did this."

Will stepped into the doorway and challenged her, "Shall we try again, Cherenka?"

He raised his pistol and fired. Steele pulled one of the men in front of her. Will's shot left a smoking hole in his chest. She drew her pistol and fired back as Will ducked behind the wall. "Saris, you're a dead man! You should have never come back."

Will mocked her, "I know you're feeling your age, Cherenka, so I thought I'd offer you an opportunity to surrender."

"Don't call me Cherenka!" She quickly fired six shots at the door jamb. The metal became molten red.

"What's wrong? You don't like your name anymore?"

"It's you I don't like." She fired several shots at the wall, three feet from the doorway. The wall's surface grew hotter.

Will pushed Shanna away and dove into the doorway. Four shots blew a hole in the wall where they had been standing. Will fired two rounds and scurried out of the way. One of his shots grazed Steele's cheek.

She ridiculed him, "You'll have to shoot better than that, Will."

"If I wanted to kill the woman in you, I'd have to look real hard."

"I'll give you something real hard!" she shouted angrily.

Will laughed. "All that time at the academy I looked up to you as a tough female officer. Now I find out you're just a pile of recycled junk."

"You rotten son of a bitch!" Steele yelled angrily. She drew a second pistol and marched toward the doorway. She fired at the walls on both sides of the doorway.

Will scrambled in one direction for cover and Shanna scrambled in the other. Shanna drew a knife and knelt low against the wall. Steele charged into the hallway and fired in both directions at eye level.

Shanna threw one of her daggers at Steele's head and ducked around the corner. It embedded inside Steele's left ear. Steele screamed in pain as she yanked at the dagger. Sparks erupted from the wound and blood trickled down her neck. She bent over and moaned as she struggled to pull the dagger out.

"Can't you take something for that headache of yours?" Will asked her. "All that moaning is killing me."

"You'll pray that I kill you swiftly," she warned. Will shot her in her left shoulder blade. She didn't even flinch.

"I'm coming for you, Steele. Wherever you go, I'll be right behind you."

Steele's comrade reached into the hallway and pulled her back into the room as Will and Shanna crept close to the door. Will burst through the doorway and scanned the room. There was no sign of Steele or her accomplice.

Shanna entered and looked about, as surprised as Will. "Where'd she go?"

"Maybe the portal was in here." He ran his hand along the near wall.

Shanna ran her hand across the opposite wall. She felt something warm and soft. Her hand disappeared through the wall. "Will! I found it." She reached further through the wall and watched her arm disappear.

Will declared confidently, "I think we finally have her on the run." He leaned shoulder first into the wall and passed smoothly into a portal. Shanna held onto his belt and followed closely. They entered a gray circular tunnel made of a soft fiber.

"This is sure strange," Shanna commented. "It's like a spider web."

"The portal's powered by a different type of energy," Will explained.

They followed the tunnel for several minutes until Shanna asked, "Could it be a trap?"

Will answered nervously, "I wondered that as well." Suddenly, he realized what was about to occur. "We've got to go back!"

"Why? What's wrong?" Shanna asked. He took Shanna's hand and yanked her arm. They raced back to the entrance of the portal. When they reached the entrance, Will shouted, "Dive!" They dove out of the portal.

Flashes of light followed a few seconds behind them. Four hot pulses of molten plasma turned the next three offices into smoking rubble.

Will breathed a sigh of relief. "Wow. That was close."

"How did you know?"

"I told her I'd be right behind her. I guess she figured that I meant it literally."

Shanna got up slowly and brushed herself off. Will rolled onto his back and took a few deep breaths. "What's wrong, Will?"

"That was way too close." Shanna helped him to his feet. Will glared at Shanna and then chastised her, "When I tell you to do something, don't ask questions. Just do it. We could have died in there." Shanna realized he was right and apologized.

Will attempted to re-enter the portal but it was gone. "How about that?"

They returned to the elevator and descended to the lower level. When the doors opened, they received their usual welcome from the soldiers. Nearby, eleven unarmed Gallian soldiers waited nervously with Xerxes.

"Looks like you've got your facility back, Xerxes. Steele is gone."

"Thanks, Will. Now what do we do?"

"You'll stay here and secure the building. The *Leviathan* is on the way to help. We're going back to Gallia. That's the only way to end this mess."

Then Will asked Xerxes, "Can you make a portal back to the *Phantom* from here?"

"Do you have the portal control unit on you?"

Will hiked his sleeve up and held his arm out. Xerxes programmed it through several buttons. "Point it at the wall and press the red button," he instructed. Will obeyed and they watched a green portal form on the wall. "Good luck. We'll be waiting for your return," said Xerxes, anxious.

Will shook Xerxes' hand. "See you soon, my friend." He then ordered the Gallians, "Come with us, guys. We're going on a trip."

Will, Shanna and the Gallian soldiers entered the portal. They walked along a sandy path tinted green and teal. Will and Shanna noticed how apprehensive the first Gallian out of the elevator was.

Will asked, "What's your name?"

"Tsig. Why do you ask?"

"I'm Will Saris and this is Shanna, my wife. If we're going to help, it makes a difference if we know a little bit about each other."

"You two are really serious about being friends."

"Of course. Why wouldn't we?"

"Cherenka has taught us that the other races look down on us as a scourge. She said no one wants to be associated with us."

"That's a total lie. I can assure you that other races have no such feelings."

"You mentioned our one-time shipping operation. My grandfather was one of the leaders of GSS. They had a great relationship with all the races until the Weevil took over."

"We believe that Steele, or Cherenka as you call her, orchestrated the raid by the Weevil on your station," Shanna said. They emerged from the portal into the *Phantom's* main quarters.

"Why don't you contact Bastille and Maya? Hopefully they'll have an update for us." Will suggested to Shanna.

Will sat at the table and noticed the Gallians were uncomfortable standing there. "Why don't you guys have a seat?" he offered.

Tsig cautiously approached the table and sat down. "Is it true what you said about Yord?" The other soldiers listened intently.

"Yes, it is. Perhaps later, we'll get you up to the surface for a tour."

"Were you serious about letting us rebuild our shipping operation out of GSS?"

"Well, someone's got to do it. Your people seemed to have a really good knack for it. Do you think you can put the program back together?"

"We have all of the data in our history block," Tsig replied.

"What's a history block?" questioned Will, now curious.

"It's like your computer, but it is historical information only. Every detail and transaction from my forefathers is documented in the block."

"That's excellent. You and your comrades already have the necessary guidance to get you back into business."

Tsig remarked cheerfully, "It would be great to live above ground again."

His statement befuddled Will. "What do you mean above ground?"

"Cherenka instructed us to move underground and to prepare for war. We built many weapons to repel invasions from the Weevil and other races that we hadn't even heard of."

"I can't believe that! You've been living underground all this time since you lost the station?"

"Yes, we have."

"That's terrible!"

Shanna returned and informed Will, "Maya's reached GSS with five Fleet battle cruisers and the *Luna C*. The *Leviathan* will be at the facility on Earth by the end of the day."

"Excellent."

"You seem to have quite a bit going on," Tsig commented.

"I have good people working with me," Will told him.

"What will you do with me and my comrades?"

"We're returning to Gallia. I want you to help me convince your boss that we can turn things around for the Gallians and explain what Cherenka is up to."

"We'll do our best. Our Major is positive that only by following Cherenka into war can we ever free ourselves and return to the surface."

"We'll have to find a way to change his mind, although I'm sure it won't be easy."

"You know him?" Tsig asked.

"Yes. We've already met and aired our differences of opinion."

Will asked Shanna, "Can you get something for our guests to eat and drink. I'll get us started on the way to Gallia."

Shanna accommodated the Gallians while Will returned to the flight deck.

· · · · · · · ·●●●●●●●●●●· · · · · · ·

The *Luna C* took a position near the front of the GSS space station. The five remaining ships were cloaked and drifted in from behind the huge portal ring. Maya opened the communication channels and waited patiently. Talia checked the scanners for missile locks from the space station and reported, "So far, so good. They aren't locking onto us with anything."

Maya operated the transmitter and called, "Come in GSS station. This is the *Luna C.*"

Talia paged downstairs on the intercom, "Mynx. Neva. You girls picking up anything?"

They sat at a comm/nav console and listened carefully to several sounds from inside the space station. Mynx retrieved the handset for the intercom and answered, "It's Mynxie. We're picking up some chatter. Nothing decipherable, though."

"Are you sensing any changes in the sound patterns?" Talia asked.

"Negative. They seem pretty relaxed down there."

"Thanks. I'll check back." Talia looked at Maya. "What do you think?"

Maya reached inside her pant leg pocket and pulled out a small spray bottle. She held it up for Talia to see.

"What's that?" Talia asked.

"Arasthmus saved me a bottle of Weevil-away." The girls chuckled.

"What's our next move?"

"I think we should go on board," Maya decided.

Talia scanned the station once more. "Looks like the middle level is empty. We could enter there."

"This is too easy," Maya uttered.

"What else can we do?"

"We have backup if we need it. We're wasting time sitting here."

"Then, let's go." They left the cabin and descended the stairs.

Mynx and Neva heard them. Mynx asked, "What's happening, Maya?"

"Talia and I are going inside GSS. We'll stay in touch."

Jack entered the main quarters and overheard her comment. "I see we're ready to make a move."

"Yes we are. Coming, dear?"

"Of course."

"Can you inform the other ships that we're going aboard?" Maya asked. "I'll be in touch shortly." Mynx immediately sent a message to the Fleet ships.

Neva affirmed, "I can track you with the extractor beam. You give the word and I'll have you out of there in a jiffy."

"Thanks," replied Maya. "I feel better about that."

"This has all the makings of a Weevil trap," Neva cautioned.

"I know, but there's absolutely no hint of activity to indicate a threat."

"I don't get it," Mynx remarked. "It makes no sense."

Maya collected three RG-7s from the locker. She handed one to Talia and one to Jack. "You know the deal. We kill with extreme prejudice."

"I've been waiting for a long time to give those punks some payback," Jack declared.

"Neva, place us on the second level in a vacant area," requested Maya.

Neva monitored the monitor as she completed a scan. "I've got just the place."

Jack and the girls stepped onto the transport platform. Maya announced, "Ready when you are, Neva."

"Here we go. Good luck, ladies. You, too, Jack." Neva activated the transporter and a bright flash filled the room then the platform was empty.

"I'd feel a lot better if I was with them," Neva uttered uneasily.

"We've got to keep an eye on things from here," Mynx reminded her. Neva glanced at the monitor. A large green field formed on the left side of the monitor. "What are you scanning?" she asked.

"The portal area. Why?" inquired Neva.

Mynx slid over and positioned herself in front of the monitor. "Oh, shit!"

"What is that?"

A huge bright light shot from the portal ring underneath the space station. Mynx cried, "We're...!"

The light struck the *Luna C* and in a flash the ship was gone.

· · · · · · ●●●●●●●●●●● · · · · ·

Maya and Talia crept down a long hallway with their pulse pistols aimed ahead of them. Jack searched in the other direction, checking each room as he went.

"Which way to the control room?" Talia whispered.

Maya pushed a button on the transmitter fastened on her wrist. "Mynx, are you there?" There was no answer.

Maya held the transmitter near her ear and still heard nothing. "Mynxie. Neva. Are you there?"

"What's wrong?" Talia asked.

"There's only static. I don't understand."

"Maybe it's the material in the shell of the station."

"No, we could monitor the Weevil from outside. There's no reason we shouldn't be able to contact the ship."

Talia shuddered and said, "Maybe we walked into their trap and the *Luna C*'s gone."

"Don't say that," Maya blurted.

Talia leaned against the wall nervously. Maya placed her hand gently on Talia's shoulder and urged, "Come on, Talia. Whatever the case, it doesn't change our mission."

Talia asked uneasily, "What could have happened to them?"

"We'll find out soon enough. Now concentrate or we'll both wind up dead."

Two Weevil in insect form burst around the corner and charged at the girls. Maya fired twice at the closest one and staggered it. She fired again and it fell to the ground dead.

Talia tried to aim her pistol and fire but her hands trembled from fear. Maya backed away and aimed at the remaining Weevil. It lunged at Talia before she could fire. Talia froze and watched in horror as it drove both its spear-shaped arms down through her shoulders and into her chest.

"No!" Maya screamed. She fired three shots and blasted the creature's head off. The creature and Talia slid to the ground. Blood seeped from Talia's mouth as she stared vacantly at Maya.

"No, Talia!" Maya burst into tears.

Jack rushed down the hall toward them and watched in shock as Talia took her last breath. Maya wept on his shoulder.

Jack asked her somberly, "Do you want to go back to the ship?"

"What ship?" Maya whimpered.

"What's that supposed to mean?"

"I'm not getting a response from my ship. Nothing at all. Just static."

Jack became frightened. This time there was no safety net and no one was coming to save them. He placed his hands on Maya's shoulders and shook her while saying, "Maya, you've got to keep it together. Both our lives depend on it."

"What can I do?" she sniveled.

"Damn it, Maya. You're an officer and a real good one. I'm not going to let you off that easy."

She regained her composure. "I'm sorry, Jack."

"Don't be sorry. I need you to be strong." Jack hugged her tightly and asked, "Now what are we going to do about this?"

"Maybe I can raise one of the other Fleet ships." She tinkered with the settings on her transmitter and spoke into the device. "Come in, *Daedalus*. Are you there?" She waited anxiously and tried again: "*Dauntless*, are you there? *Dreadnought*? Someone answer me!"

Finally a woman's voice came across the transmitter. "Maya, it's Saphoro. I'm on the *Avenger*."

"What's wrong with the *Luna C*? I can't raise her."

There was a brief silence. Maya called, "Saphoro?"

"I'm here. *The Luna C* – she's gone."

"What happened?"

"Our scanners showed some kind of super burst from the portal. They never had a chance."

"A super blast?" Jack said. "It must be some new weapon on the other side of the portal."

"Do you want us to attack?" Saphoro asked Maya.

"No. Not until I find out what caused that burst and disarm it."

"I understand. We're standing by for your orders."

"Thank you, Saphoro." Maya turned to Jack and uttered, "It looks like we're on our own, Jack. Let's cause as much chaos and confusion as possible."

"Add death and destruction to that list," quipped Jack.

They marched down the hall with their pistols aimed ahead of them. When they turned the first corner, five Weevil fired a barrage of shots at them. Maya knelt next to the wall and fired repeatedly. Jack stood tall, leaned flat against the opposite wall and fired continually as well.

One of the Weevil shots glanced off Jack's shoulder. He cried out, "Ah, you bitch!" He returned fire. Together, he and Maya killed the remaining Weevil.

Maya tended to Jack's shoulder. She peeled back his shirt and examined the wound. "It's not too bad, Honey."

"I'm good. Let's keep going." They proceeded to a gated stairwell.

Jack scurried up the steps and dashed past an open doorway. He peeked inside and saw the control room, crowded with Weevil.

Maya paused at the other side of the door. Jack whispered, "It is the control room and there's a lot of Weevil in there."

"Let's take out as many as we can. I'd like to keep one alive if possible for questioning."

"No problem."

They burst into the control room and approached the center consoles. The Weevil were focused on several monitors and failed to notice them until Maya fired five shots, inflicting mortal wounds on two of the Weevil.

Jack fired six times. He killed four Weevil and wounded two. The remaining three charged them. He fired twice more and killed two of them. Maya shot another between its beady eyes, killing it quickly.

Jack stepped over the bodies, looking for a living candidate to question. One injured Weevil struggled to get up.

"What about this one?" Jack asked.

Maya glanced at the Weevil survivor and said, "He'll do."

Jack fired three shots that destroyed another Weevil's head and shoulders. He reached down and grabbed the remaining Weevil by its mangled arm. He yanked it to its feet and shoved it into a chair.

Maya dashed around the control room, searching for other survivors. There were no more to be found. She peeked into the stairwell which was clear. She slammed the door shut and approached toward the controls. "Jack, secure the door," she ordered.

Jack replied cynically, "Yes, dear." He wrestled a large cabinet away from the wall and strained mightily as he shoved it in front of the door. Maya studied the controls, trying to determine what powered the portal.

Jack sat on the ground with his back against the cabinet. Sweat poured from his forehead and brow. "I could use your help, Jack," Maya complained.

He sighed as he stood. "Yes, dear. Right away."

"Don't be a smart-ass."

Jack stood beside her and studied the controls. Maya noticed the sweat on his forehead. "Jack, you're sweating. Are you alright?"

"I'm just honky-dory. You know that cabinet was pretty heavy."

"Yes, and you are so strong." She kissed his cheek and pressed a switch for one of the monitors. "That's one of the things I find sexy about you."

"Huh?"

"You heard me."

The monitor illuminated, displaying various graphs and charts. "What exactly are you trying to do here?" Jack asked.

"What do you think I'm doing? I'm trying to shut down the portal."

Jack chided, "Not with that panel you won't." He walked to a tall panel with several lights illuminated vertically and pointed to a small monitor. "See this?"

"Yeah. What about it?"

"The station is injecting too much power into the portal. It's also being supplemented from another source."

"Are we going to blow up?"

"No, nothing like that. The portal is on its way to becoming self-sustaining. Unfortunately, it will double in size and absorb us along the way."

Maya replied sarcastically, "Oh, so that's all? Anything else Mr. Cool?"

"Yeah. We have about three hours and ten minutes until this occurs."

"Why would they want a larger, self-sustaining portal?"

Jack went to another panel and suggested, "Let's find out."

"How do you know so much about this stuff?" Maya asked, curious.

Jack pointed at the center of the room. "Look. It's a mirror image. Half of the controls are for this side of the portal and half are for the other side."

Maya realized he was right and declared proudly, "Jack, you really are a genius!" She hurried around to the panels on the other side of the room.

Jack smiled. "Aren't you glad you have me?"

"Not now, Jack."

He frowned at her in disappointment. "Well that's gratitude for you," he grumbled.

Maya immediately operated several sections of the main console for the other side of the portal. Jack strutted over to her and watched the monitors. Data scrolled on two of the monitors. The third monitor showed three large shapes, surrounded by a multitude of smaller shapes.

Jack teased, "Come on, Maya, tell me…" Suddenly, Jack realized what he was looking at. He pressed three buttons below the monitor.

"Tell you what?"

Jack ignored her and scanned the other side of the portal. He initiated a close-up scan of the large objects. The images became magnified.

Maya stopped scrolling data and turned her attention to Jack. She noticed the intense look on his face. "What are you doing?"

Three gigantic ships were displayed on the monitor with long wands directed toward the portal. Behind them were thousands of Weevil fighters and battle cruisers.

Jack was staggered by the size of the invasion fleet. "Armageddon is just a step away," he muttered under his breath.

Maya was shocked at the three gigantic ships on the monitor. "What the hell are those things?"

"That's what destroyed the *Luna C.*"

"How the hell do we stop them?" she asked.

"You figure out where the other side of the portal is. I'll figure out how to shut it down," instructed Jack.

Maya suddenly realized that her Fleet ships were waiting in the path of the approaching force. She immediately activated the comm/nav panel and contacted them. "Listen up, commanders, this is urgent. Check in now." One by one they reported in.

"This is the *Dauntless*. We're on."

"The *Avenger* is here."

"*Daedalus*, on."

"The *Dreadnought* is on."

"*Apocalypse* is here, too."

"Get out of the area as fast as you can," Maya ordered. "The Weevil have a monstrous invasion force waiting on the other side of the portal. They have three ships with a new weapon mounted on their upper hulls. That's what took out the *Luna C.*"

Saphoro replied from the *Avenger*, "What about you and Jack?"

"We're trying to shut down the portal. If we don't do it soon, it will become self-sustaining. Then nothing can stop them from coming through and taking over this part of the galaxy."

The commanders wished Maya and Jack good luck. One by one the ships evacuated the area. Maya commented sadly, "Now it really is just us, Jack."

Jack kidded as he accessed different data banks of information, "Oh, Maya, you love this. It's the grandest ending you could have wished for. The ultimate sacrifice in front of your peers."

Maya scrolled through pages of data and information. "Jack, you're a royal pain in the ass."

"This would make a great movie. All we need is for the cavalry to come in and rescue us," teased Jack.

"Sorry, love. No cavalry in this movie."

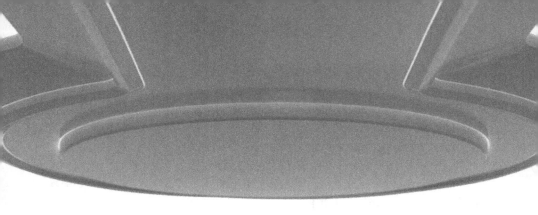

XII

A RACE
AGAINST TIME

Will studied the monitor to his left. He saw the green, glowing mass of a portal coming up. He scanned ahead and programmed the feedback into the computer.

"What's that for?" Shanna asked.

"The scanners probe the other side of the portal. The feedback tells me where the portal will take us. It's programmed into the computer which selects the quickest course for us to take."

"That's amazing."

Will watched as the data scrolled across the bottom of the monitor. "Looks like a winner."

"Any sign of Steele?"

"Not yet. I wonder if she went to GSS or straight to Gallia."

The monitor beeped three times. Will looked at the display. "Someone from the *Avenger* is calling."

Saphoro's face appeared. She looked distressed. "Will, thank goodness I got you."

"What's up, Saphoro?"

"The *Luna C* is gone."

Will's face paled. "What do you mean gone?"

"Something blasted it from the other side of the portal under GSS."

"Any survivors?"

"Maya, Jack and Talia are on board GSS. Maya ordered us out of there. No word on the others."

"Any details?"

"She said there were three huge ships with a strange new weapon mounted on their hulls. She also said the ships were part of a gigantic invasion force. It was imperative they find a way to close the portal. She said if they didn't stop it soon, it would become self-sustaining."

Will then realized what was happening. "That's their doorway to our part of the universe! They're coming with everything they've got."

"Is there anything we can do to help them?"

"Keep your distance for now. We'll try and rescue them."

"We'll be ready if you need us."

"Thanks, Saphoro." The monitor went blank. Will looked ill as he stared at the stars on the monitor.

"Do you think there's any hope?" Shanna asked him.

"There's always hope. It's just a matter of having enough of it."

Shanna sat back down in the co-pilot's seat. She stared sadly at the monitor until she fell asleep.

Will wondered what to do. He set the ship on 'auto-pilot' and went down stairs. Tsig was orating in front of his companions about how they could rebuild the shipping business. He saw Will coming and paused.

Will announced humbly, "Sorry for interrupting, fellas." He sat down at the table.

"You look troubled, Will," Tsig observed.

"Yes, I am. Maybe you or your comrades can help."

"What can we do?"

"Here's the problem. GSS has the portal ring underneath."

"That's been there since the station was built," Tsig said.

"The Weevil have found a way to make it self-sustaining. When that happens, there is a giant armada waiting on the other side of the portal to invade our part of the universe. They'll be coming soon," explained Will.

"Is that Cherenka's force?"

"I don't know. Maybe that's a surprise they have for her."

"How many ships?"

"It's massive."

"We can't match that."

"Tsig, they have three ships that are very big with strange weapons mounted on their hulls. They fired through the portal and destroyed one of my ships. Any ideas?" Tsig looked nauseous. The other Gallians seemed frightened. "What's wrong?" Will asked.

"Are you sure about this?" Tsig asked.

"My people wouldn't lie. If we can rescue them, I'm sure they'll have more information." Tsig bowed his head in shame. "What's going on, Tsig? You know something about this."

Tsig explained somberly, "We built those ships. They were supposed to be our salvation."

"What are they doing in the hands of the Weevil?"

"Someone must have gone to Gallia and commandeered them. Oh, what have we done!"

"Look. Maybe there's something we can do to stop this."

Tsig replied hysterically, "It's too late! Our people could be dead already."

"Can you contact anyone through our comm/nav system?"

"I don't know. We use a unique mode to communicate and the magnetic shell around our world distorts typical transmission signals."

Will pointed to the comm/nav set and said, "There it is if you'd like to try. Let me know if you find out anything."

"I'll do my best, Will."

Will returned to the flight deck. He considered what scenarios could have transpired that caused the Gallians to lose their three behemoth ships so easily.

Shanna stirred and awoke. "How long before we can contact GSS?" she asked.

"Another hour."

• • • • • • • • ● • • • • • • • • • •

Jack shouted across the control room, "Any luck, Maya?"

"I'm getting close. They're from the second sector in the Ursa-Omega quadrant. I don't have an exact location yet."

"Ursa-Omega. Is that like bear's ass?"

"I'm glad you can laugh at a time like this," chastised Maya.

"How else can I keep my sense of humor?"

"It would be really funny if you could make a mirror that would reflect anything they shoot at us."

Jack thought for a minute. "You know, Maya, you just gave me an idea."

"What, the mirror?"

Jack replied crazily, "Yes, a mirror! A big, stinking mirror! One that uses lots of energy."

"I think you've lost it, Jack!"

"No, I think I've just found it." He jumped around to several different panels, turning things on and off.

Maya approached him. "CCK-17," she said.

"What's that?"

"CCK-17. That's the home of the Weevil."

Jack looked at her with a stone-cold stare. "Are you sure?"

"Yes, I'm sure."

"Get me as much data as you can on CCK-17. Load it into a crypto-node."

"And do what with it?" Maya asked cynically.

"If we input that information in our data bases, everyone will know where they live, how strong they are and we'll all have a better chance to stop them."

"Not if we're dead."

"No, but we'll launch it to the other ships. To any ship. Someone has to stop them if we don't."

"Alright, I'll take care of it."

"Thank you, dear." Jack came over to Maya's side of the control room. He sat in front of a particular panel and scrolled through several pages of data. He entered numerous coordinates into the computer as he read.

"You really think you can make a mirror big enough to scare these creeps?"

"Do you doubt me?"

"Yes, I do! I'll bet you anything you can't make whatever this thing is you're trying to do."

Jack challenged her, "What do you want to bet?"

"Anything you want, Jack Fleming."

Jack grinned and echoed, "Anything?"

"Yes. I said anything."

Jack leaned across the panel until his nose nearly touched hers and said, "Marry me."

"What did you say?"

"Marry me. Maya, will you be my wife?"

Tears formed in Maya's eyes and she complained, "Jack, your timing sucks. If only we could."

Jack asked seriously, "If I get us out of here, will you marry me?"

"Of course, you fool. I wondered how long you were going to wait." She kissed him gently on the lips. "Now get us out of here."

"Maya, you just made me the happiest guy in the world." He enthusiastically resumed his task.

Maya read the elapsing time from one of the monitors. "Jack, we only have about an hour left."

"I know. I know."

Maya watched the scanners on the front side of the portal. She noticed a small green mass forming in front of the large mass.

"Jack, what's happening here?"

"I'm making a mirror."

Maya's interest grew as she watched him manipulating controls like a mad man. "You really do have an idea what you're doing, don't you?"

"Yup. See if you can contact any of your ships. Have them come back to GSS."

"What for?"

"Bait."

"Bait! You want to use my ships for bait?"

"Uh-huh."

"What if you're wrong?"

"I won't be."

Maya watched the green glowing mass grow. "How big will it be?"

"Exactly the same size as the first one."

Maya pressed the button on her transmitter and raised it near her mouth. She hesitated, then lowered it.

Jack stopped typing. "You don't believe me," he said disappointedly.

"It's not that, Jack. I can't risk their lives."

"It's not a risk. I'm sure of this."

"Where did you ever learn to make a portal let alone do something like this with it?"

"Bastille."

"Bastille? What does he know about portals?"

"What did he know about cloaking devices? What did he know about increasing engine speed on the *Phantom*?"

"But that has nothing to do with portals."

"Just remember our bet."

"Of course. I will marry you if you succeed." Jack resumed his work.

The Weevil beat on the door and interrupted the serene atmosphere. "So much for my concentration," Jack muttered.

"Keep going, Jack. I'll watch the door."

· · · · · · ● · · · · · · · · ·

The tracking system beeped, interrupting Will's tranquil mood. He sat up.

"Is it the space station?" asked Shanna.

"Sure is." He activated the monitor and turned on the transponder. "Come in, GSS. This is *Phantom*."

Maya and Jack were startled when they heard Will's voice. Maya raced to the monitor. She was elated to see Will's face. "Will! Where are you?"

"Breathing down your neck. Are you alright?"

"So far. We don't have much time."

"Don't worry. We'll get you out of there."

"No! You can't. Get away from the portal."

Will activated the cloaking system. "It's okay. They won't see us."

"No, you'll be killed."

"I got it!" Jack exclaimed.

"What's wrong with Jack?" Will asked.

"I'll let him explain." Maya resumed watching the door as the banging subsided.

Jack's face appeared on the monitor. "Hello, buddy! You're just in time."

"It's good to hear your voices. What's going on, Jack?"

"I learned a trick or two from Bastille. I've created another portal to siphon power from the first one. That should slow the self-sustaining process down. Then I designed the portal to send anything that passes through it back where it came from."

"No kidding! Are you sure it'll work?"

"Positive. The only problem is that we have just eleven minutes until both portals become self-sustaining and absorb the space station."

"I can be there. It'll be close, but I can be there."

"You'll never make it, not in the *Phantom*."

"Don't you remember what Bastille did to the engines? I'll be there in time."

"I'm counting on you, Will."

"No problem. Hey, can you do me a favor?"

"Sure."

"See if you can contact Steele on one of those ships on the other side of the portal."

"And tell her what?"

"Tell her I'm coming through. Ask her to call me."

"Are you sure about this?"

"Absolutely. I miss her metallic smile already."

Jack enjoyed the joke and ordered Maya, "Forget about the Weevil, Maya. We're getting out of here." He sent out a signal to the other side of the portal. The monitor beeped three times. "Talk about quick service," he kidded. Jack activated the receiver and waited patiently. An ugly Weevil face filled the monitor.

Maya grew nauseous. "He's much uglier than the others."

"Yeah, and bigger, too," Jack commented.

The Weevil snarled at them and said through a garbled transmitter, "I am Thorus, leader of the Weevil Empire. What is it you wish to say before you die?"

"I'm called Jack. Just a minute of your time is all I ask. Can we speak to Steele?" Thorus laughed hysterically at him. "Look, we don't have much time," Jack complained. "Can we speak to her?"

Thorus replied, "I don't think that will be possible."

"Tell her that Will Saris is on his way. Ask her to call him."

Thorus laughed again. "I don't think she's in a talkative mood." He reached off-monitor and picked up Steele's head. He set it on the table in front of him. "You can tell her yourself, but, as I said, I don't think she's in a talking mood."

"Whoa!" exclaimed Jack.

Maya looked at the monitor, appalled. "Gosh, they did her good."

Jack taunted, "How much of her could you really sink your teeth into?"

Thorus laughed again. "I like your sense of humor. It's a shame you can't stay around to share it with me."

"Just remember, I warned you about Will Saris. He'll be excited when he finds out you're in striking distance."

"Good. I have a little something for him, too."

"Well, Thorus. It was nice talking to you but I have to pack. As you know, we're leaving soon. Ta-ta." Thorus almost choked he was laughing so hard.

Jack shut off the monitor. "Hey, the Weevil aren't trying to get in anymore."

"That's because we're going to blow up. They bailed out," said Maya somberly.

Will's voice rang from the speaker over the monitor, "Hold on, children. It's play time." Jack and Maya were quickly whisked from the space station and transported aboard the *Phantom*.

Will cheered as they sped toward the space station. He turned off the cloaking system and scanned the area ahead of them. Shanna enjoyed his relaxed manner. She didn't quite understand the science of what they were doing, which made it easier for her to deal with. Maya and Jack returned to the flight deck.

"Welcome back, friends!" Will exclaimed.

"Uh, we have to talk about something," Jack said.

"Why so serious?"

"Um, Steele is dead. Thorus showed us her head."

"Who's Thorus?"

"The Weevil leader."

Will was surprised and responded happily, "No kidding?"

"Nope. He's waiting for you on the other side."

"I'm counting on it." Will paused and thought for several seconds then remarked, "Gee, I'm going to miss old Steele."

"Screw that," Jack muttered. "She deserved much worse than she got."

"Any hope for the others?"

"None," Maya told him. "Mynx and Neva were on the *Luna C*. Talia was killed by a Weevil on the space station."

"That really sucks," Will commented sadly. The cabin became quiet.

Three huge ships directed their antennae at the portal and glowed bright red. Beams of energy emitted from each and intersected in front of the portal. They formed an orb that changed to yellow and shot into the portal.

On board the *Phantom*, they watched anxiously as they approached the portal. A large ball of energy shot toward them, growing rapidly as it drew nearer.

"Jack, you'd better be right about this," Maya stammered.

"This doesn't look good," Shanna muttered. Will became nervous and broke a sweat on his forehead. Suddenly, the ball of energy was gone. The *Phantom* rocked violently and spun wildly for several minutes.

Jack and Maya were tossed across the back of the cabin. Will and Shanna hung on to their seats desperately until the ship finally stabilized.

"Wow, what a ride!" Will howled.

Jack and Maya quickly scurried to their feet. "What do you see, Will?" Jack asked excitedly.

Will studied the monitor and saw nothing. He activated the long-range scanners. "How about that? The space station is still there and everything else is gone!"

"What? The station is still there?"

"Sure looks that way."

"No enemy ships?"

"Nope. Looks like we're home free," Will announced, pleased.

They cheered and yelled joyfully. Will leaned across and kissed Shanna. Jack and Maya hugged and kissed in the doorway.

Maya said disappointedly, "Well, Jack. It looks like you won yourself a bet. I'm all yours."

Jack looked hurt, "You didn't believe in me?"

"Of course, I did. But what woman ever admits that to her man?"

"Yeah. Yeah."

"What's this all about?" Will asked.

Jack explained, "I've asked Maya to marry me and she accepted."

"But, Jack! Will teased. "What about Eve's and being single with no responsibility?"

"I could use a little responsibility and Maya's worth it," Jack confessed. "Besides we can still go to Eve's for drinks." Maya scowled at him. Jack reconsidered and said, "Well, maybe we won't go to Eve's." Will and Shanna laughed at him.

Maya explained, "You see, keeping a man is like keeping a pet. You deny him some things but you reward him with others. I deny Jack a trip to Eve's and reward him with a kiss. That's fair, isn't it?"

Will artfully avoided answering the question. Shanna helped him by asking, "Where to next?"

"Gallia."

Jack and Maya were shocked. Jack asked shakily, "Did you say Gallia?"

"Uh-huh."

Someone knocked on the door. "Can you open that, Jack?" Will asked.

"Sure. Who else is…?" Jack opened it and saw Tsig standing there. "Uh, Will."

"Yeah, Jack."

"You know there's a Gallian in the doorway."

"Yeah. That's Tsig. Tsig, this is Jack and Maya." Jack cautiously shook hands with Tsig.

"Is everything alright?" Tsig asked. "That was a rough ride and the boys got a little nervous."

"Everything is beautiful, Tsig," Will declared. "Looks like you fellas' can have your space station back."

"Really! Just like that?"

"Yup. The portal is closed and I'm taking you home. We'll see if everything is okay."

"Thank you, Will," Tsig responded and then left the cabin.

"How do you do that?" Jack asked.

Will was amused with the situation and replied innocently, "What, Jack?"

"Become friends with people who want to kill you."

Will dodged the question and suggested, "Why don't you two get cleaned up and relax for a bit. I'm sure you're exhausted."

"Good idea!" Jack exclaimed.

Maya whispered something in his ear and he promptly left the cabin. "If you don't need us, we'd like to sleep in during our stop on Gallia," Maya requested.

Will teased, "But Maya, who's going to take charge?"

"Shanna will. Ta-ta."

Will kidded, "Oh, boy. We're in trouble now."

Shanna frowned at him. "What do you think we'll find on Gallia?" she asked.

"Good question. Perhaps Steele brokered a deal to sell the new weapons to the Weevil. But, then, maybe the Weevil decided to just take them."

Tsig returned and stepped into the cabin. "I hope I'm not interrupting anything."

Will looked at him. "No, of course not. Come on in." Tsig hesitated as he stood in the doorway. "What is it, Tsig?"

"Do you have any information on Gallia yet?"

"No. We're not close enough to initiate scanners."

"I'll come get you as soon as we're within range," Shanna offered.

"That would be greatly appreciated. Thank you, Shanna." He turned to leave.

Will called out, "Tsig."

The Gallian stopped in the doorway. "Yes, Will."

"What if your people are gone? I mean if it's a reality, what will you do?"

"I don't know."

"If that happens, we'll help you. You won't be stranded."

"Thank you, again. You are a true friend." Tsig left.

"I wonder if we stemmed the Weevil attack." Will wondered.

"If we did, then Bastille is a hero," Shanna replied. "His ingeniousness has saved us once again."

"I'll give him special thanks when we return."

"What about the space station?"

"The Gallians need to take control of it. I think we can trust them to run an honest business."

"Even with the Major involved?"

"He could be a sticking point. We'll deal with that later."

"I guess Maya's going to need a new ship," surmised Shanna.

"It sure looks that way. It's a real shame we lost Talia, Mynx and Neva. They were good people."

"We'll never forget them. Now, I think I'll go take a nice hot shower," said Shanna.

"That's not fair. I'm flying."

"Come on, Will - auto-pilot! Do I have to think of everything?"

"Well, I guess we do have a ways to go before we reach Gallia." Shanna took him by the hand and led him downstairs.

XIII

THE
AFTERMATH

When Will and Shanna returned to the flight deck, Will initiated another long-range scan and sent out a beacon. Shanna sat back and watched him. He glanced at her and asked, "What are you up to?"

"Just thinking about how happy you make me."

"I'm sure I'd make you a lot happier if we weren't racing around the universe, trying to save everyone."

"I know. It's a big responsibility."

"This isn't the kind of adventure I had in mind," complained Will.

"Me neither. Maybe this will lead to something bigger for us."

"One can only hope. I'm not having fun. Tenemon and Steele are gone. I don't have any idea who this Thorus character is. The Major is just a boring creep and, for all we know, he's dead, too."

Shanna stroked the back of Will's neck. "Don't worry, we'll find you some new friends to play with."

"Stop it, Shanna. I'm not a baby."

"Yes you are. You're my baby."

"Cut it out. I hate when you do that."

"I know."

A pinging from the control panel interrupted them. Will breathed a sigh of relief. "That should be Gallia coming up."

They studied the monitor intently. A small object appeared on the upper right corner of the monitor. "That's Gallia all right."

"I'll inform Tsig," Shanna said.

Will activated a communications line and sent a second beacon toward Gallia. There was no return of his signal. He grew uneasy as he activated the short-range sensors and watched anxiously. The planet was still too far away to provide accurate details. He kept his eyes trained on the monitor as the image of Gallia grew larger and the details of the surface became more defined.

Shanna and Tsig entered the cabin. Shanna offered Tsig the co-pilot's seat and knelt between the men. Tsig was uneasy as he watched the monitor.

"Maybe their communications were disabled to keep them from calling for help," Shanna suggested.

Tsig explained, "Gallia is like a giant magnet. We can stimulate it and make it stronger to bring down alien ships. The strength of the magnet could crush a ship against the surface of the planet."

"So whoever took your ships must have been invited to the surface," Will suggested.

"That's right. Or perhaps they were lured out of the protection of the magnetic field. Then they could have been ambushed."

"But how could they get those ships so far from Gallia in such a short time?" Shanna asked.

"Perhaps they lured the ships into a portal. Then they'd be defenseless," Will answered. He scanned the planet's surface. Several of the mountainsides were destroyed and replaced by huge craters.

Will recalled firing the torpedoes earlier and thought, *I know I didn't cause all of that damage.*

Tsig looked dismayed. "Looks like they targeted all of our bases."

"Where should we start looking for survivors?" Shanna asked.

Tsig pointed to a familiar mountain range. "That was our main headquarters."

"I know it well," Will responded. Tsig looked at him with a bewildered expression. Will explained, "We visited Gallia a short while back. That's when we discovered that Cherenka was involved with your people." He pointed out a small clearing near the mountainside. "We'll land there."

Tsig was confused. "How would you land a ship this big on that tiny ledge?"

"The same way we landed without being destroyed by your magnetic fields."

Will positioned the *Phantom* a safe distance from the planet's surface. He programmed coordinates into the extractor beam control panel. "Come on, Tsig," Will directed. "Let's find out if anyone survived." He left the flight deck and descended the stairs.

Shanna assured Tsig, "He knows what he's doing."

"This method of transport is new to me."

Will gathered everyone on the transport platform. He announced, "Two red lights on that panel will illuminate three seconds before we transport," he instructed. Any questions?" No one responded.

The red lights flashed and a few seconds later they were on the surface of Gallia. The Gallians were impressed with the expediency of the transporter.

"Follow me," Tsig ordered. "Stay together in case of an ambush."

They passed through a dark tunnel and entered the remains of the transport bay. The hangar was collapsed on the gate side. Several Gallian ships were crushed under the rubble. Two ships appeared to be undamaged just beneath them. The ground was littered with the corpses of equinostriae, Weevil and Gallian corpses.

"It looks like they tried to fight," Tsig remarked somberly. Will was saddened as he surveyed the carnage.

They reached a damaged steel platform that overlooked the hangar and surveyed the area below but there was no sign of survivors.

Tsig led them to the rear of the platform. He picked up a large boulder and tossed it out of the way. Several other Gallians picked up boulders and moved them aside as well. They reached a damaged steel door that was partially crushed and bent off its hinges.

Two of the Gallians grabbed the door and pulled with all their strength. The door budged slowly at first and then broke free of the debris. They tossed it aside. It made a loud, clanging sound that echoed through the silent hangar.

The group entered another tunnel and descended a lengthy distance. At the end, the soldiers found many of their family members hidden safely within the confines of several stone rooms. They gathered together and hugged. The tunnel was filled with the sounds of sobbing voices.

"I'm so glad for them," Shanna remarked.

"Me, too," Will said.

Tsig approached them with a female Gallian and two infants. "This is my wife, Serna; my daughter, Taph, and my son, Pilst."

Will greeted them politely, "It's a pleasure to meet you."

"We're so happy you're safe," Shanna added.

Serna replied in a hoarse voice, "It's nice to meet you. I must say, I never expected humans to save us."

"Things will be much better for you," Will told her. "We'll help you in any way possible."

"Unless you can get us out of this hole, I don't think there's anything you can do."

"I believe we can do that."

Serna looked at Tsig with hope in her eyes and asked, "Is this true?"

Tsig smiled at her and replied, "Yes it is. Will's friends have recovered our old space station. They'll help us restart the shipping business that Gallian once thrived on."

"That's wonderful!" Serna cried out.

Tsig asked, "Will, who will protect us while we regroup?"

"We'll offer you an opportunity to join our alliance. As I mentioned before, Yord is returning to what it once was. Many leaders from around the sector have already pledged their allegiance to our coalition."

"Will they accept us?"

"I'm sure we'll have no trouble convincing them. Your presence makes the alliance stronger just by reestablishing your shipping operation. That will help all of us."

One of the Gallian soldiers interrupted them and shouted, "Tsig! We've found Major Duowd! He's injured."

Tsig turned to Will and Shanna. "Will you come with me and explain your offer to him?"

"Of course. I'm sure he'll be glad to see me," Will commented with a bit of sarcasm in his voice.

"Wait here Serna. This is business," Tsig advised. Serna slapped Tsig's head. Will and Shanna were shocked. Tsig returned a blow to the side of Serna's head. They head-butted each other.

"Follow me," Tsig then instructed Will and Shanna. He led them down another tunnel lit by torches on the wall.

"Tsig, what was that all about?" Shanna asked.

"What?"

"That head-slapping, head-butting stuff you and Serna just did."

"That is like the lip-smacking, tongue touching ritual you and Will did on the ship."

Will laughed at Shanna and teased, "Do you want to know more?"

"Uh, maybe I'll pass for now."

"Come on, Shanna. It might be good to try some new and creative methods of showing our love."

"Oh, stop it, Will," Shanna grumbled.

They entered a stone room. Major Duowd lay on a tiny cot. His chest cavity was ripped open and his right leg was broken nearly off. Tsig greeted the Major.

"Tsig. You've returned." The Major's voice was shallow.

"Yes, I have."

"The Tridents have been taken from us. Cherenka set us up."

"I know about it."

The Major began to tremble in anger. "The whore turned our ships against us. She destroyed our military capacity through deceit."

Tsig replied coolly, "We have defeated the enemy."

"Which enemy? We have several."

"Cherenka is dead. The Weevil have been repelled from the space station and from the sector."

"How did you know?"

"We had some help from our friends."

Will and Shanna stepped into the room. "Hello, Major," Will greeted him.

Major Duowd's eyes filled with surprise. "Saris! What are you doing here?"

Tsig explained, "Will is the reason we are triumphant today."

"So I was wrong about you," Duowd muttered.

"All you had to do was listen," said Will.

"Lift me up, Tsig."

"You can't get up, sir. You're in bad shape."

"I'm dying, you fool. Lift me up." Tsig did as he was ordered. "Help me to the assembly area."

Tsig wrapped an arm around Duowd's waist and escorted him up the tunnel. Will and Shanna followed. They returned to the area where Tsig's family waited. There were a hundred or so other family members there, accompanied by their soldier kin.

Major Duowd shouted with every ounce of energy he could muster, "Listen up, everyone." The group became silent. Soldiers closed in to hear their leader's words.

"I don't have long to live so I'll keep this short. Tsig will be my successor. Follow and obey him as you would me. Give him your support and make Gallia proud of her sons. Thank you, everyone." He coughed violently and ordered, "Set me down, Tsig."

Tsig gently set him on the ground. "Promise me you'll not make the same mistakes I did," requested the Major.

"You have my word, sir."

The Major lowered his head and passed away. Tsig called out, "A warrior leaves us!"

The soldiers responded in unison, "May he reach his destiny. Hoo! Hoo! Hoo!"

"I'm sorry for your loss, Tsig."

"Thank you, Will. That means a lot."

"What is your first objective?" Will asked.

"We need to make GSS operable again for our purpose. Can you take me and ten of my men to the station?"

"Sure."

Tsig approached his wife. "Serna, I will be back soon," He then called out, "Simian! Come forward."

A young Gallian soldier approached and stood at attention. "Yes, sir!"

"I want you to take charge of establishing a settlement on the surface. You'll need to protect our people from the equinostriae as well."

The soldier was surprised by the order and asked, "Sir?"

"You heard me. We're done living underground in fear. The surface of Gallia is ours again!" Cheers greeted Tsig's words.

Tsig hand-picked ten men capable of restoring GSS to a shipping hub. He turned to Will. "We're ready."

"Then let's get going."

They departed the area and ascended the tunnel to the outside ledge. A short while later, they were aboard the *Phantom*.

· · · · · · · · ●●●●●●●●●●●● · · · · · · · ·

Will and Shanna were seated in the flight deck of the *Phantom*. Will programmed the coordinates for the space station while Shanna set the long-range sensors.

"Do you think that it's over, Will?" she asked.

"No. It's never over. Just a lull."

"Why don't you get some sleep?" she suggested. "I'll keep an eye on things."

"You know, Shanna, I can't remember the last time I had a good sleep."

"Well, now's a good time."

Will set the seat horizontally and slept.

Later on, Shanna heard a beeping sound from the control panel. She checked the sensor monitor and saw a large green mass forming in the center. To the right was the space station. "Who else could possibly make a portal like this?" she wondered. Suddenly a large blip emerged from the portal. "Oh, no! This can't be good." She urgently shook Will.

He quickly sat up. "What's wrong, Shanna?"

"Look at the monitor!"

Will set the seat upright and studied the monitor. "Did you hear any sounds from the console?"

"Yes. It beeped six times."

"That's bad. That's real bad."

"What does it mean?"

"It means someone already has a lock on us." He watched the monitor and was awed by the size of the blip. "That's one big ship!"

"What do we do?"

"We're turning around!" He quickly changed the ship's direction. The monitor showed a yellow wave extending from the blip toward the *Phantom*.

"Great! Now what?"

Tsig burst into the cabin. "What's going on? We changed directions."

"We've got a problem," Will told him. The *Phantom* was drawn toward the blip on the monitor.

Tsig looked at the monitor and was horrified. "It's one of the Tridents!"

"What's that mean, Tsig?"

"We're caught in a magnetic tractor beam."

"They already have a weapons lock' on us."

Tsig bowed his head and said somberly, "We're finished."

"There's got to be something we can do."

"You can't escape it once it's locked on to you."

"Is this one of your ships, Tsig?"

"Yes."

"How did they make another portal?"

"The three Tridents have a power capability similar to the *Leviathan*. They can make their own portals but only for a short while. It takes a long time to restore full power after using it like this. That's why the *Leviathan* is so valuable. That's why Cherenka tried to get it for us."

"Why? To copy its power generation system?"

"Absolutely. It's the perfect system. It replaces what it uses at exactly the same rate."

"All this time we've been sitting on something this special and we had no idea."

"Maybe that's where Bastille's ideas and improvements came from," Shanna suggested.

"Well, I'll be. Wait until I see Bastille," Will uttered.

The monitor beeped three times. "I guess we're about to find out who's behind this." Will keyed a button and watched the monitor intently.

Thorus' face appeared. "Well, now. If it isn't my good friend, Will Saris. I thought we'd have met sooner."

Will replied cynically, "Damn, Thorus, looks like you're a lot uglier than your peers."

"That wasn't nice what you did to my ships."

Will thought back to the sequence of events. "Ah, yes. It was like firing into a mirror, wasn't it?"

"You cost me two Tridents and a slew of battle cruisers. I owe you for that."

"That's alright. I'm sure you have more."

Tsig whispered a warning to Will, "If you know any tricks or miracles, now is a good time to use them."

Thorus asked, "How is it that you've managed to make a living of cheating death? We thought we had you a number of times."

"I guess I'm just one lucky guy." He nervously fingered his belt. He felt the strings from the brown, leather sack. He suddenly remembered what Xerxes told him about its contents.

Thorus complained, "I guess if you want something done right, you have to do it yourself."

"Maybe I should share my secret with you," suggested Will.

"Humor me, Saris, just one more time before I rid the universe of your miserable body."

Will reached into the sack and removed the golden skull. He stood up in the cabin and looked forward. On one monitor he saw the Trident approaching. "This skull is my lucky charm. I'm going to show you what I mean."

"What are you doing, Will?" Shanna asked, alarmed.

"Don't worry. I have everything under control, I think."

"No, you don't," Tsig muttered. "We're doomed."

Will wrapped his hands in the sack. Then he placed them around the skull and held it in the direction of the Trident. "Go ahead, Thorus. I'm ready when you are."

Thorus was entertained and remarked, "I see you still have your sense of humor, even in the face of death."

"I hate to see you go, Thorus. I don't have many friends to engage in playful conversation anymore. They're all gone."

Thorus erupted in unbridled laughter. "Don't worry, you'll be joining them soon enough."

Tsig warned frantically, "Here it comes!"

Shanna embraced Will as he closed his eyes and prayed that the skull really worked like Xerxes said it would. A moment of silence passed.

Will paused and wondered, *Am I still alive?* He opened his eyes and looked at the sensor monitor. He was stunned.

The Trident was a ball of yellow and red plasma until it disintegrated into nothing. "It worked! The Trident's gone!" Shanna cried out.

Tsig stared at the monitor in amazement. "How did you do that, Will?"

Will nervously lowered the skull and replied cynically, "I told him this was my lucky charm." He initiated another scan. The monitor was blank except for GSS. "Well, Tsig, looks like you're going to make it to your old space station."

Tsig hugged him. "You saved us, Will! You did it!" His big frame and strength difference hurt Will, who gasped and pleaded, "Please, Tsig. You're crushing me to death."

Tsig realized what he was doing and set Will down. "Sorry about that."

Will quickly covered his head. Tsig asked, "What was that for?"

"So you don't slap or head-butt me."

Tsig and Shanna laughed hysterically. Will finally put his arms down and asked, "What's so funny?" Tsig pointed at Will and laughed harder. Will became annoyed and demanded, "Come on, Tsig. What's so funny?"

Shanna explained, "The slapping is for mating purposes only."

Will was embarrassed and muttered, "Oh, I get it." He programmed the coordinates for the transport control panel. "Come on, Tsig. You guys have to go."

"What's the hurry?"

"I've got a lot of things to take care of and a new daughter back home on Yord."

Shanna and Will accompanied the Gallians down the stairs. Tsig ordered his men onto the platform as Will waited by the local transporter controls.

Maya and Jack emerged from the cabin at the end of the hall with surprised looks on their faces. "What's going on now, Will?" Jack asked.

"Back to bed, you two," Will ordered. There was a flash on the platform and the Gallians were gone.

"What happened? Did we miss something?" Maya asked.

"Not a thing," Will said. "Just a boring, routine trip back to Yord. Oh, Thorus said hello." Jack and Maya's eyes widened in fear.

Will said calmly, "Don't worry. He's out of the picture now."

"What in the world is going on?" Jack asked anxiously.

"Can the two of you keep an eye on things upstairs for a while? I'll fill you in later."

"Sure," Jack replied. "Are you alright?"

"Just fine."

"So what happened?"

"I'll tell you later. It's all good now." Will took Shanna by the hand and led her to their cabin.

"I think they've lost it," Maya remarked.

"You're not kidding."

· · · · · · · ·●● ● ● · · · · · · · ·

Jack and Maya reviewed the data log on the flight deck. "Looks like everything must have worked out on Gallia. They came, they saw, and they left." Jack said.

Maya picked up the skull and the brown leather sack from the floor. "I wonder what Will was doing with this."

"You never know with him."

Will entered the flight deck, beaming happily. "Did I hear my name?"

Jack kidded, "Well, well. Look who's back. It's been almost thirteen hours in Earth time. It's a shame you shut down all those portals. We're taking the long way home."

"Tsig will reprogram the portals to make sure we don't have any more surprise visits for a while."

"I hope you feel better after your rest," Maya remarked.

"Oh, I feel much better."

Shanna entered the cabin with a frown on her face.

"What's wrong, Shanna?" Maya asked.

Will answered, "Shanna wanted to engage in some Gallian mating techniques." He slapped her butt and chuckled.

Shanna yelped, "Ouch! That hurts, you jerk."

"Will, what did you do to her?" Maya questioned.

He replied slyly, "I showed her a mating trick of my own."

Jack looked at Will in surprise and said, "You really did that?"

"Uh-huh."

"Wow! That's…" Maya glared at Jack. "…not nice," he finished.

Will replied defensively, "Shanna wasn't complaining at the time."

"When is your next lesson, Shanna?" Jack asked.

"I'm done with Gallian tricks. From now on, I'll just be a boring, submissive wife."

Will rubbed her butt and said, "Oh, darling, you're never boring."

She looked at him with puppy dog eyes. "Really, Will?"

"Really." He gently pushed Shanna through the doorway.

"Where are you going?" Maya asked.

"Back to bed." He closed the door behind him.

"There's something seriously wrong with them," Maya complained.

Jack latched the door shut and dimmed the lighting. "What are you doing?" she asked, curious.

Jack pulled her onto his lap and kissed her. Maya replied coyly, "Oh, I see."

· · · · · · · · · ● · · · · · · · · · ·

Jack dialed in the codes for the *Leviathan* and sent a beacon out. Maya knelt on the floor and rested her head on his lap. When Will and Shanna returned, Will asked anxiously, "Did you contact the *Leviathan* yet?" Just then the monitor pinged three times.

"Of course. That's them now," Jack replied. He pressed the 'receive' button.

Bastille's face appeared on the monitor. "Well, hello friends. How are you doing?"

"Hey, Bastille! How come you never told me about the *Leviathan's* power generation system?"

"What about it?"

"That it's perfect!"

"Oh, yeah. I didn't think you cared."

"Well, everyone wanted the *Leviathan* not for its weaponry but because of its ability to generate exactly the amount of power it uses."

"That's great, isn't it?" Will frowned at him. Bastille promised, "I'll tell you more when I see you."

"Yes, you will," Will replied sorely.

"Anything interesting happen out there?"

"Yeah. The Gallians built three Tridents. Furey helped the Weevil steal them. They turned on Steele and killed her. The Tridents destroyed the

Luna C. Tahlia, Mynx and Neva are gone. We almost got vaporized as well by the remaining Trident."

Bastille looked surprised. "The Tridents, huh? And you escaped their fabled death ray?"

"Not exactly. I used it against them."

"Now you have to explain that one to me, my friend."

"How are things on Earth?" Will asked, concerned.

"We regained control of the Washington area and a perimeter of seven hundred miles. After taking out a number of tactical targets, I think we stemmed the tide of anarchy in this region for a while. Xerxes' guys appear to have things under control."

"That's good news."

"Do you want us to remain here?"

"No, come on home. Tell Xerxes that the skull worked. He'll understand."

Celine pushed Bastille out of the way and hollered, "I'm coming back to the *Phantom*. You guys are having all the fun." They heard her cackle in the background as the monitor went blank. Will sat in the co-pilot's seat and put his feet up on the console. Finally, they could relax and enjoy their new boring role as King and Queen of Yord.

XIV

THE HOME FRONT

A week later, the leaders of Yord's allies met at the palace to discuss the recent series of events. Will and Jack entered the hall and proceeded to the head table. Shanna stood behind the table, holding Marina in her arms. She was dressed elegantly in a teal robe and a tiara of gold on her head. Her blond hair draped down over her shoulders.

Will couldn't help but think how beautiful she was. The perfect mother dressed so regally. *But then there's her other side*, he thought. *Black leather, knives; her zest for passion.*

Maya and Serna stood on Shanna's right. Will and Jack took note of the modified Fleet uniform worn by Maya. The front of the shirt had a deep-V neckline. The pants were tight, black leather. It was hardly befitting the conservative Fleet style. She wore the bandages around her head like a badge of courage which served as a reminder that their enemies were not to be taken lightly.

"Looks like Maya's getting some tips on battle garb from Shanna," Jack commented.

"Our girls are looking good," replied Will.

"Yes, they are."

Mariel marched across the room and paused in front of the table. "I'd like to ask for a moment of silence for several friends lost in the recent battle. Everyone bowed their heads in reverence. Finally, Mariel raised her head and announced enthusiastically, "Introducing the King and Queen of Yord, Will and Shanna." The attendees cheered.

Will quipped, "Just like old times, huh, Shanna?"

"Not quite. Why don't you hold your daughter?" she responded and craftily handed Marina to Will, while smiling radiantly at everyone. Jack poked Will in the side in jest.

"Thank you all for attending this most important meeting," Shanna announced. "We've overcome a significant crisis which taxed the strength of our new union. Before I go on, we want to thank Emperor Rethus and Empress Atilena for providing significant military support for Yord during our mission; for the bold support from General Asheroff of the Boromian race; and the tactical support from Queen Siphra of Attrades despite her time of need." Everyone in the room cheered.

Shanna continued, "I'd like to give thanks to our commander, Maya, of the former Space Fleet. She represents the new Union Fleet which consists of all allies of Yord. The Space Fleet is no more as we knew it." The room remained silent as the attendees waited for further explanation. "I present Maya to detail the new changes."

Maya took center stage and smiled at Jack. "Today begins a new phase in the rebuilding of Yord and her allies. GSS was recently controlled by a Weevil insurgent group under the name of Galactic Security Services. Fortunately, they have been ousted from the quadrant. The Gallians have taken control of the station for the purpose of restoring their most vital business operation of shipping and transportation. The correct title of the station has been returned, and slightly modified – Gallian Space Services. They will be conducting business effective today, with main hubs on Yord, on Earth and on Alpha-17. Here, today, representing the Gallians is Lady Serna, wife of the Gallian Emperor Tsig." Serna waved to the audience.

Maya continued, "I've dispatched Union ships to GSS to ensure their operation continues safely. I've also dispatched ships to Earth to restore law and order. We are currently investigating the Weevil invasion force to find out what remaining accesses they might have to our quadrant. This is necessary to prevent further aggression into our territory by these predators. Thank you." She glanced at Jack, again with a beaming smile.

"Are there any questions?" Shanna asked. No one spoke.

Will announced sadly, "We have been through a tough time. We've lost some close friends as well. Forgive us if we seem distracted from our duties at this time. We need time to heal and recover from our losses."

A representative from Taurus-21 stood and asked giddily, "Is there any truth to the rumor that the King was injured during Gallian mating rituals with the Queen?" Everyone chuckled.

Will was quite embarrassed. "Tough one, buddy," Jack whispered.

"Where did that come from?" Will whispered back.

Mariel stepped in and chastised the man. "It is very inappropriate to ask a question of that nature to anyone of royal stature."

Shanna interrupted. "I wish to address that question. We have among us a variety of races, cultures, languages and habits. We must be open to accepting others as we want them to accept us. To answer your question, do you think the King would still be standing if we engaged in Gallian mating rituals?" Everyone laughed and cheered her response.

Will whispered to Jack, "Now it's my turn." He handed Marina to Shanna.

"You go, boy," Jack whispered.

Will paced the room. "It is good to kid around and show humor from time to time, but we need to realize there are entities out there that would do anything to undermine what we are building. This recent mission was accomplished by the active involvement of leaders in the alliance directly. Rethus, Tsig, Sarbis, Tachemon and many others who personally manned their ships in defense of Yord. Because of their support, the Weevil invasion was deterred. Because of them, we were in a position to cripple that invasion force with minimal casualties to our people. We overcame several cunning enemies who had us at an extreme disadvantage. This is a huge victory for the Union. So, please keep in mind that there are still serious threats out there and that there are capable people fighting for your safety and well-being. Thank you all." He smiled at Shanna and took Marina from her.

"Well done," Jack said.

Shanna placed her arms around Will and kissed him zealously. Everyone in the room stood and applauded them. Will thanked them, once more. Shanna bowed and waved as she and Will departed.

As Will hurried from the room, Jack attempted to follow him but Maya grabbed him subtly by the belt. "There is one more thing I'd like to mention," she announced to the group. "Jack Fleming has asked me to marry him and I've accepted." The room rang with cheers. "We're having a small ceremony so as not to detract from the victory celebration beginning this evening. So, please, wish us lots of luck." The guests clapped for them.

"What was that for?" Jack asked, embarrassed.

"You did ask me to marry you, didn't you?"

"But I wanted to make the announcement."

"I told you before, when I get something, you get something. Have I ever disappointed you?"

"I guess not." She hooked an arm in his and led him through the hall. She answered questions and discussed the mission while Jack waited impatiently in silence.

·······●●●●●●●●●●●●·······

Will and Shanna strolled down the hall with Marina. Keira met them and asked, "Would you like me to take Marina so you can mingle with your guests?"

"That would be great. Thank you, Keira." Will handed Marina to her.

Shanna followed Will to the bedroom. He closed the door and gazed lovingly at her. He became sad and hugged her tightly.

"What's wrong?" she asked, surprised.

Will sat on the bed. "They're gone and they're not coming back."

"I'm going to miss them, too," Shanna mentioned to him.

"This is what I feared most and it's happened."

Mariel knocked then opened the door. "I'm sorry about your friends," she commented. "It's a tragic loss."

"Thanks, Mariel. It's going to be empty without them," Will uttered sadly.

"Just think how close you and the others came to joining them," she reminded him. "You've overcome great odds in your success."

"But was it worth it?" Will asked, looking disappointed.

Mariel placed a hand on his shoulder and emphasized, "Look what you have around you, Will: A wife and daughter, friends, a new world in Yord and a rebirth of your people as well. I'd say it's a fair trade-off."

"Then I think we'll try and enjoy our home life for a little while before we plan any more adventures."

Shanna noticed that Will looked distracted. She inquired, "Do you think Thorus is gone?"

"For some reason, I don't think so."

"Who's Thorus?" asked Mariel.

Shanna replied uneasily, "Will's new friend." Mariel looked frightened and left the room.

"Maybe me and Thorus can work things out?" kidded Will.

Shanna pushed him back on the bed and said, "Forget about Thorus. I want you to spend some quality time with me."

Will chuckled. "Here we go again."